AF834296

# Power Systems

Electrical power has been the technological foundation of industrial societies for many years. Although the systems designed to provide and apply electrical energy have reached a high degree of maturity, unforeseen problems are constantly encountered, necessitating the design of more efficient and reliable systems based on novel technologies. The book series Power Systems is aimed at providing detailed, accurate and sound technical information about these new developments in electrical power engineering. It includes topics on power generation, storage and transmission as well as electrical machines. The monographs and advanced textbooks in this series address researchers, lecturers, industrial engineers and senior students in electrical engineering.

**Power Systems is indexed in Scopus**

Oscar Danilo Montoya Giraldo •
Walter Julián Gil-González •
Alejandro Garcés Ruiz

# Optimal Power Flow Methods for DC Networks

## A Convex Optimization Approach

Springer

Oscar Danilo Montoya Giraldo
Electrical Engineering
Universidad Distrital Francisco
José de Caldas
Bogotá D.C., Colombia

Alejandro Garcés Ruiz
Electric Power Engineering
Universidad Tecnológica de Pereira
Pereira, Risaralda, Colombia

Walter Julián Gil-González
Electric Power Engineering
Universidad Tecnológica de Pereira
Pereira, Risaralda, Colombia

ISSN 1612-1287 ISSN 1860-4676 (electronic)
Power Systems
ISBN 978-3-032-05733-4 ISBN 978-3-032-05734-1 (eBook)
https://doi.org/10.1007/978-3-032-05734-1

© The Editor(s) (if applicable) and The Author(s), under exclusive license to Springer Nature Switzerland AG 2026

This work is subject to copyright. All rights are solely and exclusively licensed by the Publisher, whether the whole or part of the material is concerned, specifically the rights of translation, reprinting, reuse of illustrations, recitation, broadcasting, reproduction on microfilms or in any other physical way, and transmission or information storage and retrieval, electronic adaptation, computer software, or by similar or dissimilar methodology now known or hereafter developed.
The use of general descriptive names, registered names, trademarks, service marks, etc. in this publication does not imply, even in the absence of a specific statement, that such names are exempt from the relevant protective laws and regulations and therefore free for general use.
The publisher, the authors and the editors are safe to assume that the advice and information in this book are believed to be true and accurate at the date of publication. Neither the publisher nor the authors or the editors give a warranty, expressed or implied, with respect to the material contained herein or for any errors or omissions that may have been made. The publisher remains neutral with regard to jurisdictional claims in published maps and institutional affiliations.

This Springer imprint is published by the registered company Springer Nature Switzerland AG
The registered company address is: Gewerbestrasse 11, 6330 Cham, Switzerland

If disposing of this product, please recycle the paper.

# Preface

This book compiles the most recent research on optimal power flow (OPF) studies for direct current (DC) networks, providing a comprehensive overview of the topic and offering practical insights for both researchers and practitioners. A main contribution of this work lies in the implementation of various OPF algorithms using Julia, an interpreted high-performance programming language well suited for technical and scientific computing, and JuMP, an extensive library of optimization tools available in Julia. Hence, this book enables readers to explore state-of-the-art approaches to solving OPF problems with practical emphasis.

Designed to be a valuable resource for undergraduate and graduate students in engineering, as well as for professionals in the power systems field, the book covers key topics including power flow and optimal power flow analyses in monopolar and bipolar DC networks. It provides a deep dive into both exact nonlinear programming formulations and various convex approximations that simplify these complex problems while preserving solution quality. Throughout the book, theoretical concepts are complemented by practical examples, and all numerical results are obtained using Julia software, showcasing its capability to handle a wide range of optimization tasks in power systems.

Readers will benefit not only from the detailed explanations of the algorithms and methodologies but also from hands-on experience through the implementation of these techniques in Julia. This approach makes the book particularly relevant for those seeking to deepen their understanding of OPF in DC networks, and for anyone interested in applying advanced optimization techniques to modern power system challenges.

**Acknowledgments** This book was produced within the framework of the research project: *Desarrollo de una metodología de gestión eficiente de potencia reactiva en sistemas de distribución de media tensión empleando modelos de programación no lineal*, registered with the Research Unit of the Faculty of Engineering at Universidad Distrital Francisco José de Caldas. Internal code 33787724. In addition, this book was produced within the framework of the research project: *Desarrollo de una metodología de control secundario para microrredes de corriente continua aisladas empleando control predictivo basado en el modelo*, registered with the Vice-Rectory for Research, Innovation and Extension at Universidad Tecnológica de Pereira. Internal code 6-24-9. Finally, the authors acknowledge the support provided by the Thematic Network 723RT0150 *Red para la integración a gran escala de energías renovables en sistemas eléctricos (RIBIERSE-CYTED)* financed by the call for Thematic Networks of the CYTED (Ibero-American Program of Science and Technology for Development) for 2022.

| | |
|---|---|
| Bogotá D.C., Colombia | Oscar Danilo Montoya Giraldo |
| Pereira, Risaralda, Colombia | Walter Julián Gil-González |
| Pereira, Risaralda, Colombia | Alejandro Garcés Ruiz |

# Contents

# Chapter 1
# A Review of DC Networks

**Abstract** This chapter provides a comprehensive overview of direct current (DC) networks, encompassing both low-power and high-power applications within a unified framework. Low-power applications include DC microgrids—localized distribution systems that integrate renewable energy sources, energy storage devices, and controllable loads—commonly found in residential areas, commercial buildings, university campuses, and electric vehicle charging stations. On the other end of the spectrum, high-power applications are represented by multiterminal high-voltage direct current (MT-HVDC) transmission systems, which are essential components in the development of super-grids and international energy hubs. The chapter emphasizes the growing relevance of DC technology in modern power systems due to its efficiency, flexibility, and suitability for integrating distributed energy resources. Additionally, this chapter introduces the DC power flow problem as the foundational model for analyzing the steady-state operation of such networks. While simpler than its AC counterpart due to the absence of reactive power and frequency dynamics, the DC power flow problem remains nonlinear and poses significant challenges when incorporated into optimization frameworks. This motivates the study and development of specialized solution techniques, as explored throughout the book. The formulation of the power flow problem, its mathematical structure, and its role as a constraint in various optimal power flow (OPF) formulations are discussed in detail. By establishing the theoretical and practical context, this chapter lays the groundwork for the advanced methodologies presented in the subsequent chapters.

**Keywords** DC power systems · DC microgrids · High-voltage direct current · DC power flow analysis · Optimal power flow

## 1.1 Introduction

Electric power systems have transformed the economy, culture, and even the landscape of entire countries. It is the largest and most complex machine built by human beings. In its origins, at the end of the nineteenth century, the electric power

© The Author(s), under exclusive license to Springer Nature Switzerland AG 2026 
O. D. Montoya Giraldo et al., *Optimal Power Flow Methods for DC Networks*,
Power Systems, https://doi.org/10.1007/978-3-032-05734-1_1

system used the direct-current (DC) technology, strongly defended by Tomas Alva Edison. Very soon, it was replaced by the alternating current (AC) technology, which demonstrated to be more efficient. Hence, the twentieth century witnessed a great expansion of AC electrical systems around the world. Two devices were relevant for this victory of the AC over the DC technology: the electric transformer and the synchronous machine. The former is the most suitable way to increase the voltage to transmission levels. The latter is the most efficient way of generating electric power from conventional energy resources.

Nevertheless, the twenty-first century is witnessing a rebirth of DC networks. Several portions of the grid are being replaced by this technology at low-, medium-, and high-voltage applications. At low voltage, DC microgrids and nano-grids constitute the archetypical applications. These are used to integrate renewable power resources and energy storage devices for residential and commercial users. Small industries and university campuses may also be a DC microgrid. Medium-voltage applications include data center, DC distribution networks, and energy communities [1]. Likewise, multiterminal high-voltage direct current transmission (MT-HVDC) is the most general form of DC network at high power. This approach is the key component in the concept of super-grid and energy hubs [2].

A DC grid has clear advantages in terms of efficiency and controllability due to the lack of reactive power or frequency control [3]. It is possible to transmit power for long distances without requirements of reactive power compensation. The lack of reactive power is also convenient in low-power applications (see [4] for a complete overview of the challenges of reactive power in AC microgrids). In addition, it may be an option to integrate renewable resources such as solar photovoltaics and hydrogen production from electrolysis, as well as storage devices such as batteries and supercapacitors, which are inherently DC [5]. Even some AC technologies such as wind energy may be integrated directly into a DC microgrid due to the requirement of a power electronic converter.

Indeed, the power electronic converter is to a DC network as the synchronous machine is to an AC grid. A power electronic converter is a component that works as interphase between AC and DC networks. It is also a component that enables controlling electric variables in the DC networks. Therefore, the control objective of each power electronic converter must be included into the load flow analysis as presented in this book. Below, we present a brief description of DC microgrids and multiterminal HVDC transmission. After that, the power flow and optimal power flow problems are briefly presented.

## 1.2 DC Microgrids

A DC microgrid is a localized power distribution system that uses DC technology to interconnect and distribute electricity among various resources. Said resources include renewable power generation, loads, and energy storage devices [6]. Unlike traditional AC grids, which are widely used in power distribution, DC microgrids

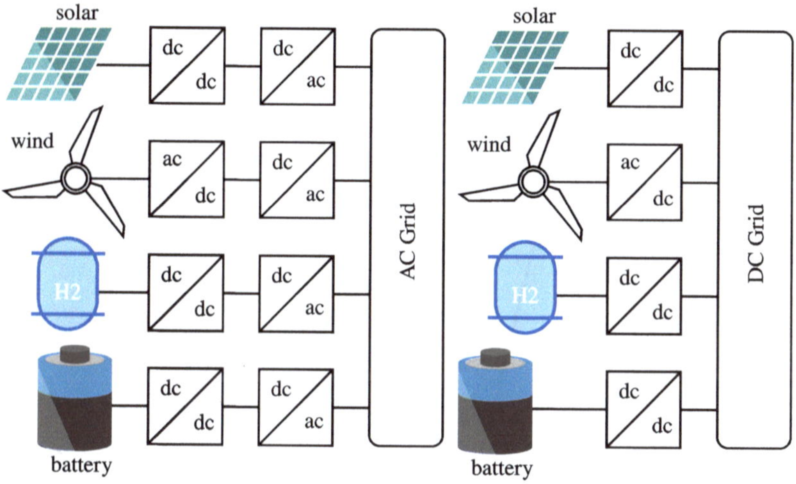


**Fig. 1.1** Example of an AC and a DC microgrid integrating different renewable energy resources. Notice the DC microgrid requires less conversion stages

operate entirely on DC power, making them more efficient for certain applications and environments. This type of grid is common in data centers [7], DC distribution [8], and electric vehicles charging stations [9] and to supply energy to remote areas [10].

This type of grid is a natural choice for integrating renewable energy resources and energy storage devices when the primary resource is DC. For example, solar photovoltaic systems, hydrogen production from electrolysis, and batteries are DC technologies that can be easily integrated into a DC microgrid. Figure 1.1 shows a comparison between an AC and a DC microgrid for integrating the same resources. Notice that the DC microgrid requires fewer stages of conversion, improving efficiency and reducing investment costs.

## 1.3 Multiterminal HVDC Transmission

DC technology has coexisted with the conventional AC grids for more than a half a century. The first commercial high-voltage direct current (HVDC) line went into service in 1954, in Gotland, on the Swedish east coast [11]. Thereafter, the HVDC technology has evolved, increasing its power capacity and ability to provide auxiliary services. Modern HVDC lines are based on forced commutated converters, such as the pulse-width modulated voltage source converter [12] and the modular multilevel converter [13]. These converters enable active and reactive power control and, in some cases, voltage and frequency control.

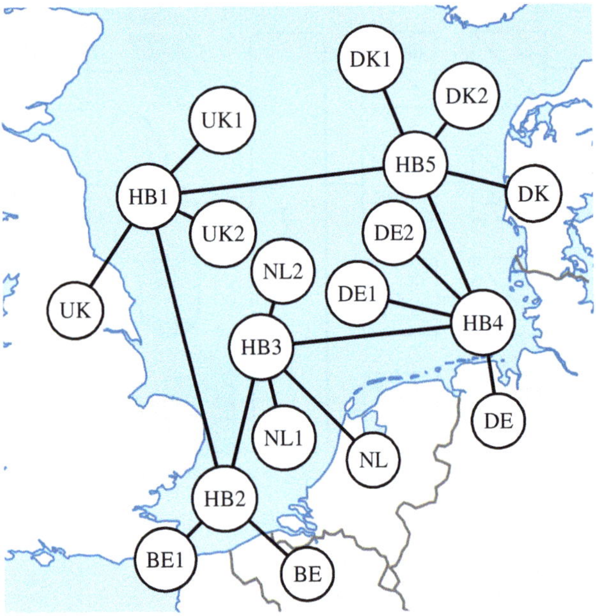


**Fig. 1.2** Meshed multiterminal high-voltage direct current transmission network with five energy hubs in the North Sea. Adapted from [2]

The complexity of HVDC networks increases as technology evolves from point-to-point transmission lines to multiterminal HVDC (MT-HVDC) transmission. Space to build new transmission lines is becoming increasingly scarce, so parts of the system must be built or replaced by DC, since the latter take up less space. Furthermore, the integration of offshore wind farms and the necessity for interconnection at country levels leads to the use of MT-HVDC and the one depicted in Fig. 1.2. This realistic test system considers the integration of several existing and planned offshore wind farms in the North Sea and the interconnection with on-land transmission networks in Denmark, the United-Kingdom, Belgium, the Netherlands, and Germany.

The size of multiterminal HVDC networks is still reduced, so its mathematical modeling is similar to the mathematical modeling of a DC microgrid, at least, in stationary state [14]. This leads to proposing a unified methodology of stationary state operation and optimal load flow for both low- and high-voltage applications.

## 1.4 The Power Flow Problem in DC Networks

The power flow is one of the most classical problems in electrical engineering; this problem has been continuously studied since the 1960s in AC networks with many variations. Recent research has been oriented to theoretical guarantees of convergence [15] and the introduction of power electronics and control devices. The main idea of the power flow problem is to determine the state of the system, providing information about loads and controlled variables in power generators. The problem can also be formulated in DC networks.

The power flow problem in DC networks is undoubtedly easier than its counterpart in AC. However, the problem is still nonlinear and induces a non-convex set into optimization problems. Therefore, it is a relevant problem for research and development. The injected power for each node $k$ of a grid is given by

$$p_k = \sum_{m \in \mathbf{N}} g_{km} v_k v_m, \tag{1.1}$$

where $v_k$ and $v_m$ are nodal voltages, and $g_{km}$ represents the nodal admittance matrix. Despite being a nonlinear equation, it has some structure that may be used for developing tailored algorithms, as presented in this book. Conic approximations and sequential linearizations can be used to obtain an approximated solution to several optimization problems under the general framework of optimal power flow.

Indeed, the optimal power flow (OPF) is a family of optimization problems which include the power flow equations as constraints. Instances of this problem are defined by the objective function and/or additional constraints. The most common instance consists in minimizing power losses subject to the power flow equations and box constraints in voltages and power flows. Other objective functions include minimizing costs, maximizing reliability or stability indices, or minimizing congestions. These objectives are usually linear or convex quadratic functions, which are easy to manage. The main challenge is given in the equations associated with the power flow in (1.1).

## 1.5 Outline of the Book

This book presents a comprehensive and structured analysis of the OPF problem in DC networks, addressing both theoretical and practical aspects through a series of specialized chapters. Chapter 2 introduces the general formulation of the OPF problem using a nodal representation, highlighting the challenges posed by constant power loads and extending the formulation to the multiperiod case. Following this, a recursive convex reformulation based on product approximation is presented in Chap. 3, employing a multivariable Taylor expansion to linearize nonlinear constraints and transform the problem into a tractable quadratic programming model. Chapter 4 introduces a hyperbolic approximation of the power flow equations, providing a simplified and computationally efficient methodology supported by recursive refinement. The next contribution (see Chap. 5) explores a recursive conic approximation that models constant power loads using auxiliary variables and convex constraints, facilitating iterative convergence to an optimal solution. Building upon this, Chap. 6 applies semidefinite programming (SDP) to the OPF problem in MT-HVDC systems, showcasing the benefits of convex relaxation for ensuring global optimality and computational efficiency. Chapters 7 and 8 present two additional second-order cone programming (SOCP) approaches, each offering distinct strategies to convexify the original non-convex OPF problem while demonstrating strong numerical performance and scalability through simulation results. Finally, the book concludes with Chap. 9 dedicated to quasi-static time-series OPF analysis in DC networks with energy storage, where a dual-decomposition algorithm is proposed to tackle the inherent large-scale and time-coupled optimization challenges. Together, these contributions provide a robust foundation for understanding and solving the OPF problem in modern DC power systems, offering valuable insights for both academic research and practical applications.

## References

1. Mutluri, R. B., & Saxena, D. (2024). A comprehensive overview and future prospectives of networked microgrids for emerging power systems. *Smart Grids and Sustainable Energy, 9*(2), 45.
2. Rodrigues, S., Pinto, R. T., Bauer, P., & Pierik, J. (2013). Optimal power flow control of VSC-based multiterminal dc network for offshore wind integration in the North Sea. *IEEE Journal of Emerging and Selected Topics in Power Electronics, 1*(4), 260–268.
3. Garcés, A. (Ed.). (2021). *Modeling, Operation, and Analysis of DC Grids*. Academic Press.
4. Mohammed, A., Sakr, E. K., Abo-Adma, M., & Elazab, R. (2024). A comprehensive review of advancements and challenges in reactive power planning for microgrids. *Energy Informatics, 7*(1), 63.
5. Albasheri, M. A., Bouchhida, O., Soufi, Y., & Cherifi, A. (2024). Enhanced supervisor energy management technique of dc microgrid-based PV/wind/battery/SC. *Electrical Engineering, 107*(2), 2285–2296.
6. Li, Q., Zhao, F., Zhuang, L., Wang, Q., & Wu, C. (2023). Research on the control strategy of dc microgrids with distributed energy storage. *Scientific Reports, 13*(1), 20622.
7. Ramirez-Marin, S.-A., Garcés-Ruiz, A., Cortés-Borray, A. F., Perez-Basante, A., & Rodríguez-Seco, J.-E. (2025). Model-predictive control with admittance matrix estimation for the optimal power sharing in isolated dc microgrids. *Electric Power Systems Research, 241*, 111380.
8. Rodriguez-Gil, J. A., Mojica-Nava, E., Vargas-Medina, D., Arevalo-Castiblanco, M. F., Cortes, C. A., Rivera, S., & Cortes-Romero, J. (2024). Energy management system in networked microgrids: An overview. *Energy Systems*.
9. Arya, H., & Das, M. (2023). Fast charging station for electric vehicles based on dc microgrid. *IEEE Journal of Emerging and Selected Topics in Industrial Electronics, 4*(4), 1204–1212.
10. Saha, S., Bhattacharjee, A., Elangovan, D., & Arunkumar, G. (2017). Dc microgrid system for rural electrification. In *2017 International Conference on Energy, Communication, Data Analytics and Soft Computing (ICECDS)* (pp. 307–313).
11. Lescale, V. F. (1998). Modern HVDC: state of the art and development trends. In *POWERCON '98. 1998 International Conference on Power System Technology. Proceedings (Cat. No.98EX151)* (vol. 1, pp. 446–450).
12. Acha, E., Kazemtabrizi, B., & Castro, L. M. (2013). A new VSC-HVDC model for power flows using the Newton-Raphson method. *IEEE Transactions on Power Systems, 28*(3), 2602–2612.
13. Hammons, T. J., Lescale, V. F., Uecker, K., Haeusler, M., Retzmann, D., Staschus, K., & Lepy, S. (2012). State of the art in ultrahigh-voltage transmission. *Proceedings of the IEEE, 100*(2), 360–390.

14. Sanchez, S., Garcés, A., Bergna-Diaz, G., & Tedeschi, E. (2019). Dynamics and stability of meshed multiterminal HVDC networks. *IEEE Transactions on Power Systems, 34*(3), 1824–1833.
15. Garcés, A., Montoya, O. D., & Gil-González, W. (2022). Power flow in bipolar dc distribution networks considering current limits. *IEEE Transactions on Power Systems, 37*(5), 4098–4101.

# Chapter 2
# The Optimal Power Flow Problem

**Abstract** This chapter presents a detailed formulation of the optimal power flow (OPF) problem in monopolar DC networks, adopting a nodal representation grounded in graph theory and circuit analysis. The chapter begins by introducing the mathematical foundations of DC power systems, including the construction of the nodal conductance matrix and the formulation of the power flow equations. A representative example is used to highlight the nonlinearities and challenges introduced by constant power loads, particularly in the context of the power balance constraints. The chapter then explores how the power flow problem can be reformulated as a convex optimization model using a potential function approach, offering a geometrical interpretation that facilitates analysis and solution. Additionally, a derivative-free iterative method based on successive approximations is developed to solve the power flow equations efficiently. Building upon this foundation, the chapter extends the OPF formulation to a multiperiod framework, incorporating time-varying variables such as generation, load, and network constraints. This generalized model includes power balance, current flow limits, voltage regulation, and device capability constraints, culminating in a comprehensive multiperiod optimization problem aimed at minimizing total energy losses. Finally, a classification of existing solution techniques is presented, emphasizing the role of convex optimization methods—both direct and indirect—as robust and mathematically grounded alternatives to heuristic approaches for solving the OPF problem in DC networks.

**Keywords** DC optimal power flow · Convex optimization · Constant power loads · Multiperiod modeling · DC network constraints

## 2.1 Grid Representation

A DC network can be represented using graph theory, similar to the representation of conventional AC networks. However, the primary distinction lies in the domain in which the model is defined: While AC networks are modeled in the complex domain, DC networks are entirely defined within the real domain [1]. This

© The Author(s), under exclusive license to Springer Nature Switzerland AG 2026 
O. D. Montoya Giraldo et al., *Optimal Power Flow Methods for DC Networks*, Power Systems, https://doi.org/10.1007/978-3-032-05734-1_2

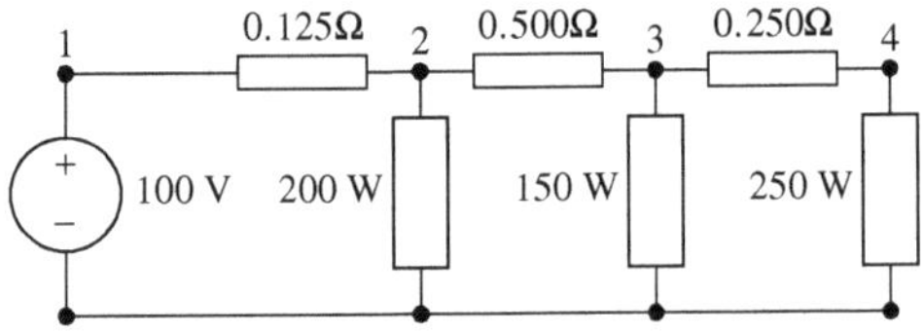


**Fig. 2.1** A monopolar DC network composed of a constant voltage source and three constant power terminals

distinction is significant, as many power flow equations in the complex domain are nonanalytical (i.e., non-holomorphic), which complicates their mathematical handling. The following representation is general and applicable to any monopolar DC network [2].

The relation between voltages and currents is given in Eq. (2.1) [3]:

$$p_k = v_k i_k, \ \forall k \in \mathbf{N}, \tag{2.1}$$

where $p_k$ is the net power injection at node $k$, calculated as the difference between the power injected by generators and the power demanded by constant power loads; $v_k$ is the voltage at node $k$, and $i_k$ represents the total current injection at the same node; and $\mathbf{N}$ represents a set that contains all the nodes except the ground where the voltage is zero. The net current injection is represented as a function of the nodal conductance matrix $G = [g_{km}]$ as given below:

$$\begin{bmatrix} i_1 \\ i_2 \\ \vdots \\ i_k \\ \vdots \\ i_n \end{bmatrix} = \begin{bmatrix} g_{11} & g_{12} & \cdots & g_{1k} & \cdots & g_{1n} \\ g_{21} & g_{22} & \cdots & g_{2k} & \vdots & g_{2n} \\ \vdots & \vdots & \ddots & \vdots & \ddots & \vdots \\ g_{k1} & g_{k2} & \cdots & g_{kk} & \cdots & g_{kn} \\ \vdots & \vdots & \ddots & \vdots & \ddots & \vdots \\ g_{n1} & g_{n2} & \cdots & g_{nk} & \cdots & g_{nn} \end{bmatrix} \begin{bmatrix} v_1 \\ v_2 \\ \vdots \\ v_k \\ \vdots \\ v_n \end{bmatrix}, \tag{2.2}$$

where $n$ is the cardinality of the set $\mathbf{N}$ (i.e., the number of elements of the set) and $G$ is the nodal conductance matrix. This equation can also be represented term by term as follows:

$$i_k = g_{k1}v_1 + g_{k2}v_2 + \cdots + g_{kk}v_k + \cdots + g_{kn}v_n = \sum_{m \in \mathbf{N}} g_{km}v_m, \ \forall k \in \mathbf{N}. \tag{2.3}$$

The nodal conductance matrix has several properties that come directly from the graph theory. For example, it is a positive definite and weakly chained diagonally dominant matrix if the graph is connected [4].

**Example 2.1** Let us consider a small numerical example with four nodes depicted in Fig. 2.1.

A base voltage of 100 V and a power base of 100 W are employed to obtain the values in per unit. Hence, the conductance matrix is given below:

$$G = \begin{bmatrix} 800 & -800 & 0 & 0 \\ -800 & 1000 & -200 & 0 \\ 0 & -200 & 600 & -400 \\ 0 & 0 & -400 & 400 \end{bmatrix}. \tag{2.4}$$

This matrix can be easily calculated in Julia using the following script:

```
using LinearAlgebra
using DataFrames
# conductance matrix
function Conductance(Lines,num_lines,num_nodes)
    G = zeros(num_nodes,num_nodes)
    for line in eachrow(Lines)
        gkm = 1/line.rkm
        G[line.k,line.k] +=  gkm
        G[line.k,line.m] += -gkm
        G[line.m,line.k] += -gkm
        G[line.m,line.m] +=  gkm
    end
    return G
end
# main functions
# input data
Lines = DataFrame( k = [1,2,3],
                   m = [2,3,4],
                   rkm = [0.125,0.5,0.250])
num_lines = size(Lines,1)
num_nodes = maximum([Lines.k;Lines.m])
# per unit representation
Vbase = 100; Pbase = 100; Rbase = Vbase^2/Pbase;
Lines.rkm = Lines.rkm/Rbase;
G = Conductance(Lines,num_lines,num_nodes)
display(G)
```

Notice this code is general for any grid stored in the data frame labeled as `Lines`.

---

Once the nodal conductance matrix is calculated, a general representation of monopolar DC networks is obtained by substituting (2.3) into (2.1), obtaining

$$p_k = v_k \sum_{m \in \mathbf{N}} g_{km} v_m, \ \forall k \in \mathbf{N}. \tag{2.5}$$

This expression corresponds to a set of nonlinear algebraic equations that require numerical methods to be solved [5]. This is known as the *power flow problem*. However, if some values of $p_k$ are unknown, then the problem allows an optimization that corresponds to the *optimal power flow problem*. The nodes in the grid can be separated among the nodes where the voltage is directly controlled and the nodes where it is not. Usually, there is only one node with controlled voltage, labeled as the slack node. In such a case, the algebraic set is reduced to $n - 1$ equations with an equal number of unknown variables. Still, this reduced algebraic set requires accurate and efficient numerical methods to obtain a solution. These methods can be classified into two groups, namely derivative-free and derivative-based approaches [6]. It is also possible to solve the problem as an optimization problem, as presented in the next section.

There is an alternative formulation to represent the power flow problem in monopolar DC networks, which is based on the branch formulation, i.e., a formulation that includes the current flow in all the distribution lines. However, it only applies to strictly radial distribution grids [2].

## 2.2 Power Flow Problem as an Optimization Problem

Although there exist several numerical methods for efficiently solving the power flow problem. However, there is an alternative approach, based on optimization, that reveals interesting geometrical properties of the model (see [3]). Let us rewrite Eq. (2.3) as follows:

$$-\frac{p_k}{v_k} + \sum_{m \in \mathbf{N}} g_{km} v_m = 0, \ \forall k \in \mathbf{N} - \{s\}, \tag{2.6}$$

where $s$ is the slack node.

Equation (2.6) is a gradient vector field that comes from a potential function $f : \mathbb{R}^n \to \mathbb{R}$ presented in (2.7):

$$f(v) = -\sum_{k \in \mathbf{N}} p_k \log(v_k) + \frac{1}{2} \sum_{k \in \mathbf{N}} \sum_{m \in \mathbf{N}} g_{km} v_k v_m, \ \forall k \in \mathbf{N}. \tag{2.7}$$

Notice that $G = [g_{km}]$ is positive semidefinite for any circuit and positive definite if the graph that represents the grid is connected. The Hessian of $f$ defines its geometric properties, namely

$$H_f = \text{diag}\left(\frac{p_k}{v_k^2}\right) + G. \tag{2.8}$$

Clearly, $H_f \succeq 0$ if $p_k \geq 0$ and $v_k > 0$. Moreover, the function is strictly convex if $p_k > 0$ and the graph is connected. Therefore, solving (2.6) is equivalent to the following convex optimization model:

$$\begin{aligned} \min z_v &= f(v), \\ \text{s.t:} \quad & v_k > 0, \ \forall k \in \mathbf{N} - \{s\} \\ & v_s = 1, \end{aligned} \tag{2.9}$$

where $z_v$ is the objective function minimized for the voltage variables in $v$. The constraint $v_k > 0$ holds automatically if the grid is not in an extreme condition such as short circuit. In that case, the solution of the optimization model (2.9) can be obtained by calculating gradient as it is an unconstrained problem:

$$\frac{\partial f(v)}{\partial v_k} = 0, \ \forall k \in \mathbf{N} - \{s\}, \tag{2.10}$$

which effectively produces the set of constraints (2.6), i.e., the set of equations regarding the power flow problem in monopolar DC networks. Notice that the function $f(v)$ is strictly convex in the case where the graph is connected. In such a case, the optimal solution is unique. This is a distinctive property of DC grids that is not general for AC grids.

**Example 2.2** The power flow problem resulting from the DC distribution network presented in Example 2.1 can be solved as an optimization problem using Julia. The following code solves the problem:

```
using JuMP
using Ipopt

N = 1:num_nodes
Nodes = DataFrame(p = [0.0,200.0,150.0,250.0])
Nodes.p = Nodes.p/p_base
model = Model(Ipopt.Optimizer)
@variable(model, v[1:num_nodes] >= 0)
@constraint(model,v[1]==1)
p_n = -Nodes.p
f_loss = 1/2*v'*G*v
@NLobjective(model, Min, f_loss-sum(p_n[k]*log(v[k]) for k in N))
optimize!(model)
v_n = [value(v[k]) for k in N]
p_n[1] = v_n[1]*G[1,:]'*v_n
display(v_n)
```

It is important to check that the Hessian of the potential function is positive semidefinite to guarantee it is a minimum point. Notice that, in a more general case, the problem may not be convex. In that case, there is no guarantee of a solution.

---

Solving the power flow as an optimization model may not be efficient in practice. In the current literature, multiple numerical methods are proposed to address this problem. A complete list of numerical methods can be found in [7]. Below, we present numerical methods that can be easily implemented in Julia.

## 2.3 Successive Approximation

A simple and efficient way to solve the power flow problem is by the successive approximation method. Equation (2.5) can be divided into two main subsets of nonlinear algebraic equations, associated with the slack node and the second one with demands, namely

$$p_k = v_k \sum_{m \in \mathbf{N}} g_{km} v_m, \quad \forall k \in \mathbf{S}, \tag{2.11}$$

$$-p_k = v_k \sum_{m \in \mathbf{N}} g_{km} v_m, \quad \forall k \in \mathbf{D}, \tag{2.12}$$

where **S** is the set that contains the slack node, and **D** is the set that contains all the demand nodes, such that $\mathbf{N} = \mathbf{S} \cup \mathbf{D}$. All the voltages in the set **D** are unknown. Once the set of Eqs. (2.12) is solved, then the value of the slack power generation in (2.11) is obtained by evaluating the voltage values in the demand nodes.

Here, we present a power flow solution approach from the family of derivative-free methods, named the successive approximation power flow method, which was inspired by the classical Gauss-Seidel method. In this case, all voltages are calculated using a matrix formulation, as given in Eq. (2.13).

$$-p_d = \text{diag}(v_d)\left[G_{ds} v_s + G_{dd} v_d\right], \tag{2.13}$$

where $v_d$ is a vector that contains all the voltage variables in the demand nodes, $v_s$ is a vector that contains the voltage inputs in the nodes with voltage control; $p_d$ is a vector containing all the demand nodes; $G_{ds}$ is a rectangular matrix that contains all the conductive effects between the slack and demand nodes, and $G_{dd}$ is a square non-singular matrix; and $\text{diag}(v_d)$ is an operation that allows becoming a vector into a diagonal matrix.

Considering the properties of $G_{dd}$, it is possible to obtain a general formula to represent the power flow problem where voltages are a function of the constant power consumption, as presented in Eq. (2.14)

$$v_d = -G_{dd}^{-1}\left[\text{diag}(v_d)^{-1} p_d + G_{ds} v_s\right]. \tag{2.14}$$

To have a solution for the nonlinear relation among voltages in Eq. (2.14), an iterative solution approach is proposed by introducing an iterative counter $t$, as follows:

$$v_d^{t+1} = -G_{dd}^{-1}\left[\text{diag}\left(v_d^t\right)^{-1} p_d + G_{ds} v_s\right], \tag{2.15}$$

where $v_d^t$ are the voltages at iteration $t$ which allows obtaining the new values for the voltages at iteration $t+1$. The following stopping criterion is used to define the convergence of the successive approximation power flow (2.15) to the power flow solution.

$$\max\left\{\left\|v_d^{t+1} - v_d^t\right\|\right\} \leq \varepsilon, \tag{2.16}$$

where $\varepsilon$ is the desired convergence error.

**Example 2.3 (Power Flow Problem)** We use the results of Example 2.1 to illustrate the application of the successive approximation method to a monopolar DC network with multiple constant power terminals. The slack node corresponds to node 1, and the rest of the voltages are given by the vector $v_d = [v_2\ v_3\ v_4]^\top$. The general successive approximation power flow formula (2.16) takes the following form:

$$\begin{bmatrix} v_2^{t+1} \\ v_3^{t+1} \\ v_4^{t+1} \end{bmatrix} = \begin{bmatrix} 1000 & -200 & 0 \\ -200 & 600 & -400 \\ 0 & -400 & 400 \end{bmatrix}^{-1} \left[ \begin{bmatrix} v_2^t & 0 & 0 \\ 0 & v_3^t & 0 \\ 0 & 0 & v_4^t \end{bmatrix}^{-1} \begin{bmatrix} 2.0 \\ 1.5 \\ 2.5 \end{bmatrix} + \begin{bmatrix} -800 \\ 0 \\ 0 \end{bmatrix} v_1 \right]. \tag{2.17}$$

We consider $v_1 = 1.0$ in per-unit representation and initial conditions $v_d = [v_2^0\ v_3^0\ v_4^0]^\top = [1\ 1\ 1]^\top$. Then, the iterative solution of Eq. (2.17) is presented in Table 2.1 for a convergence error of $1 \times 10^{-6}$.

Numerical results in Table 2.1 reveal that after five iterations the desired convergence is reached with a value of $3.35277 \times 10^{-8}$. The final voltage values at nodes 2, 3, and 4 are 99.231 V, 97.164 V, and 96.516 V.

Once the value of voltage profiles has been found, the total power losses can be determined by the following quadratic form:

$$p_{\text{loss}} = v_d^\top G_{\text{bus}} v_d.$$

The total power losses were 14.949 W. To implement the successive approximation power flow method in monopolar DC networks, the Julia programming

**Table 2.1** Iterative solution for the power flow problem using the successive approximation method in the four-node test system

| $t$ | $v_1$ (pu) | $v_2$ (pu) | $v_3$ (pu) | $v_4$ (pu) | Error |
|---|---|---|---|---|---|
| 1 | 1 | 0.99250 | 0.97250 | 0.96625 | $3.37500 \times 10^{-02}$ |
| 2 | 1 | 0.99232 | 0.97167 | 0.96520 | $1.04806 \times 10^{-03}$ |
| 3 | 1 | 0.99231 | 0.97164 | 0.96517 | $3.32744 \times 10^{-05}$ |
| 4 | 1 | 0.99231 | 0.97164 | 0.96517 | $1.05626 \times 10^{-06}$ |
| 5 | 1 | 0.99231 | 0.97164 | 0.96516 | $3.35277 \times 10^{-08}$ |

environment is used. For this purpose, three functions are created besides the functions previously created in Example 2.1.

```
function LossCal(Gbus,V)
    Ploss = transpose(V)*(Gbus*V);
    return Ploss
end
function SuccApprox(Gbus,Nodes,error,itermax)
    Gdd = Gbus[2:end,2:end]; Rdd = inv(Gdd); Gds = Gbus[1,2:end];
    Vs = Nodes[1,2]; V0 = Nodes[2:end,2]; Pd = Nodes[2:end,3];
    for t = 1:itermax
            Vd = -Rdd*(inv(LinearAlgebra.diagm(V0))*Pd + Gds*Vs);
        if maximum(abs.(abs.(Vd) - abs.(V0)))< error
            global V = [Vs;Vd];
        else
            V0 = Vd;
        end
    end
    return V;
end
```

We explain the use of each of these functions below:

i. `Conductance(Lines,L,N)` is a function that permits calculating the conductance matrix by using as inputs: `Lines` that corresponds to the information of the monopolar DC branches, i.e., the sending and receiving nodes $k$ and $m$ and the resistance value between both nodes $r_{km}$, and `L` and `N` that are constant numbers associated with the number of branches and nodes that composes the monopolar DC grid structure.
ii. `LossCal(Gbus,V)` is a function that defines the total grid power losses for a particular load condition, which is calculated using the conductance matrix `Gbus` and the final voltage profiles `V`.
iii. `SuccApprox(Gbus,Nodes,error,itermax)` corresponds to the implementation of the successive approximation power flow formula defined in (2.15), including the convergence criterion in (2.16), where `error` represents the maximum convergence error $\varepsilon$, and `itermax` means the maximum number of iterations $t_{\max}$.

The reader is invited to implement the proposed power flow algorithm in Julia to verify the numerical reports reached for the four-node test feeder.

---

## 2.4 The Optimal Power Flow Problem

The problem of the optimal power flow (OPF), contrary to the conventional power problem, corresponds to an optimization problem where the main idea is to know the amount of power injection in dispersed generation sources to minimize a particular performance indicator [8]. The OPF formulation is based on the power flow problem by introducing new variables and constraints associated with the devices' capacities and the grid code. Here, we present each one of the constraints by explaining its meaning.

### 2.4.1 Power Balance Constraint

A key difference with the conventional power flow problem is the inclusion of the dispersed generation at each node; in addition, in the case of solar sources owing to the time dependence of these sources based on the weather conditions, the time variable is also introduced. The general power balance equation takes the following form:

$$p^{s}_{k,h} - p^{dg}_{k,h} - p^{d}_{k,h} = v_{k,h} \sum_{m \in \mathbf{N}} g_{km} v_{m,h}, \ \{\forall k \in \mathbf{N}, \ \forall h \in \mathbf{H}\}, \tag{2.18}$$

where $p^{s}_{k,h}$ and $p^{dg}_{k,h}$ mean the power injections in the slack and dispersed generation sources connected to the DC network at the period $h$, $p^{d}_{k,h}$ corresponds to the constant power consumption in the demand nodes, and $v_{k,h}$ and $v_{m,h}$ are the voltage variables at nodes $k$ and $m$ at the time $h$. Observe that $\mathbf{H}$ represents a set that contains all the analysis periods, typically defined in hours or minutes.

### 2.4.2 Current Flow Calculation and Bounding

The current flow $i_{km,h}$ in a branch $km$ can be calculated as

$$i_{km,h} = \frac{v_{k,h} - v_{m,h}}{r_{km}}, \ \{\forall km \in \mathbf{L}, \ \forall h \in \mathbf{H}\}, \tag{2.19}$$

where $r_{km}$ is the resistance associated with the conductor used to interconnect the nodes $k$ and $m$. Note that the current flow is defined in a new set called $\mathbf{L}$, which contains all the branches of the monopolar DC network. To ensure the proper operation of the conductor, the total current flowing in the route $km$ must be less than or equal to the conductor's thermal bound $i^{\max}_{km}$, i.e.,

$$\left|i_{km,h}\right| \leq i^{\max}_{km}. \ \{\forall km \in \mathbf{L}, \ \forall h \in \mathbf{H}\}. \tag{2.20}$$

### 2.4.3 Devices' Capabilities and Voltage Regulation

To ensure the proper operation of the generation sources as happened the conductor capabilities, the total power generation in these devices are upper and lower bounded by its nominal rates, as defined below:

$$p_{k,h}^{s,\min} \leq p_{k,h}^{s} \leq p_{k,h}^{s,\max}, \ \{\forall k \in \mathbf{N}, \ \forall h \in \mathbf{H}\} \tag{2.21}$$

$$p_{k,h}^{dg,\min} \leq p_{k,h}^{dg} \leq p_{k,h}^{dg,\max}, \ \{\forall k \in \mathbf{N}, \ \forall h \in \mathbf{H}\}, \tag{2.22}$$

being $p_{k,h}^{s,\min}$ and $p_{k,h}^{s,\min}$ the minimum and maximum power generation limits in the slack source, and $p_{k,h}^{dg,\min}$ and $p_{k,h}^{dg,\max}$ mean the lower and upper generation bounds for the dispersed generation.

On the other hand, regulatory entities require the fulfillment of additional constraints regarding the quality of the electrical service provided to all end users. In the case of steady-state analysis, the typical condition is associated with voltage regulation in all network nodes [9]. The general voltage regulation constraint is presented in the inequality condition defined in the following equation:

$$v_{k}^{\min} \leq v_{k,h} \leq v_{k}^{\max}, \ \{\forall k \in \mathbf{N}, \ \forall h \in \mathbf{H}\}, \tag{2.23}$$

where $v_{k}^{\min}$ and $v_{k}^{\max}$ are the minimum and maximum voltage regulation bounds assigned for the distribution grid. Note that these limits are not time-dependent since normative does not discriminate periods for voltage regulation impositions.

In addition, in the OPF problem, to ensure the proper operation of the distribution system, the slack source is set with a constant voltage input, i.e.,

$$v_{k,h} = v_{\text{nom}}, \ \{\forall k \in \mathbf{S}, \ \forall h \in \mathbf{H}\}, \tag{2.24}$$

where $v_{\text{nom}}$ is the nominal operative voltage of the distribution grid.

### 2.4.4 Objective Function Formulation

The OPF problem is typically characterized by an objective function associated with minimizing the total grid power/losses energy losses [10]. An electrical distribution network's power/energy losses are the net difference between the power injection (energy provided by generation sources) and the total power consumption in constant power terminals [11]. This objective function is formulated in the equation

$$E_{\text{loss}} = \sum_{k \in \mathbf{N}} \sum_{h \in \mathbf{H}} \left( p_{k,h}^{s} - p_{k,h}^{dg} - p_{k,h}^{d} \right) \Delta_h, \tag{2.25}$$

where $E_{\text{loss}}$ is the energy losses value, and $\Delta_h$ is the length of the period where all the electrical variables are assumed as constants. In literature, two alternative forms are used to represent the objective function (2.25), which are defined in equations give below:

$$E_{\text{loss}} = \sum_{k\in\mathbf{N}} \sum_{m\in\mathbf{N}} \sum_{h\in\mathbf{H}} v_{k,h} g_{k,m} v_{m.h} \Delta_h, \tag{2.26}$$

$$E_{\text{loss}} = \sum_{km\in\mathbf{L}} r_{km} i_{km,h}^2 \Delta_h. \tag{2.27}$$

**Remark 2.1** All the ways of calculating the objective function associated with energy losses in Eqs. (2.25), (2.26), and (2.27) are entirely equivalent, with the main property that all of them are convex functions (properties will be presented in the next chapter), which will become convex optimization tools excellent alternatives to address the OPF problem in monopolar DC networks [12].

### *2.4.5 Complete Optimization Model*

To have a complete representation of the OPF model for monopolar DC distribution networks, the set of equations inequality constraints (2.18)–(2.27) is presented compactly in the optimization model (2.28).

**Multiperiod OPF Model**

$$\text{O.F.:} \quad \min\left\{E_{\text{loss}}\right\} = \sum_{k\in\mathbf{N}} \sum_{m\in\mathbf{N}} \sum_{h\in\mathbf{H}} v_{k,h} g_{k,m} v_{m.h} \Delta_h \tag{2.28}$$

st:

$$\begin{aligned}
& p_{k,h}^{s} - p_{k,h}^{dg} - p_{k,h}^{d} = \sum_{m\in\mathbf{N}} g_{km} z_{km,h}, && \{\forall k \in \mathbf{N},\ \forall h \in \mathbf{H}\} \\
& r_{km} i_{km,h} = v_{k,h} - v_{m,h}, && \{\forall km \in \mathbf{L},\ \forall h \in \mathbf{H}\} \\
& \left|i_{km,h}\right| \leq i_{km}^{\max}, && \{\forall km \in \mathbf{L},\ \forall h \in \mathbf{H}\} \\
& p_{k,h}^{s,\min} \leq p_{k,h}^{s} \leq p_{k,h}^{s,\max}, && \{\forall k \in \mathbf{N},\ \forall h \in \mathbf{H}\} \\
& p_{k,h}^{dg,\min} \leq p_{k,h}^{dg} \leq p_{k,h}^{dg,\max}, && \{\forall k \in \mathbf{N},\ \forall h \in \mathbf{H}\} \\
& v_{k}^{\min} \leq v_{k,h} \leq v_{k}^{\max}, && \{\forall k \in \mathbf{N},\ \forall h \in \mathbf{H}\} \\
& v_{k,h} = v_{\text{nom}}, && \{\forall k \in \mathbf{S},\ \forall h \in \mathbf{H}\}.
\end{aligned}$$

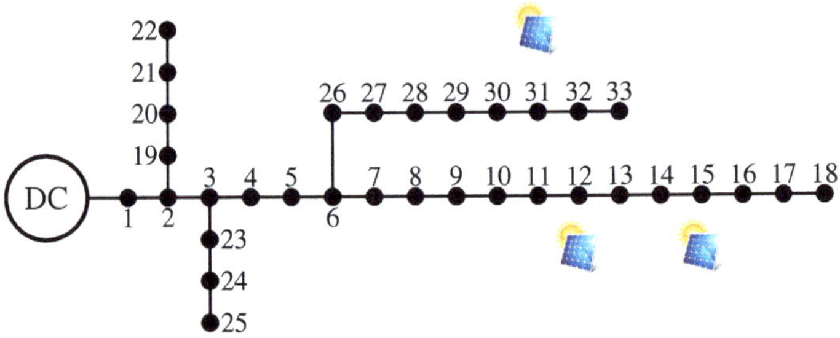


**Fig. 2.2** Monopolar DC version of the IEEE 33-bus grid

**Remark 2.2** The main characteristic of the multiperiod OPF model (2.28) is that it corresponds to non-convex optimization problem owing to the presence of the product between voltage variables in the power balance constraint [13]. This non-convexity motivates this research book with the idea of presenting multiple convex approximations to deal with OPF problems in monopolar DC networks and their potential applications in distribution system operation.

**Example 2.4** To illustrate the formulation of the OPF problem in monopolar DC networks, let us consider a generic DC network with renewable generation depicted in Fig. 2.2. This network corresponds to the DC version of the IEEE 33-bus grid proposed in [5].

---

## 2.5 Solution Techniques

The solution of the OPF problem for single- and multiperiod analysis has been widely addressed in the specialized literature with multiple optimization tools. Figure 2.3 presents the possible solution methodologies to deal with the OPF problem in monopolar DC networks.

From Fig. 2.3 it is possible to note that:

i. The solution methodologies based on metaheuristic optimization algorithms work via master-slave (or leader-follower) approaches, where the master stage defines the power generation in conventional and renewable energy by maintaining its maximum and minimum values using a metaheuristic algorithm [14]. In the slave stage, the power generations are included as inputs in the power flow problem which is solved with any numerical method, i.e., Gauss-Seidel [2], Newton-Raphson [15], or the successive approximation approach [16], among others.
ii. The convex optimization approaches can be classified into direct and indirect optimization methods. In the case of direct methods, the second-order cone

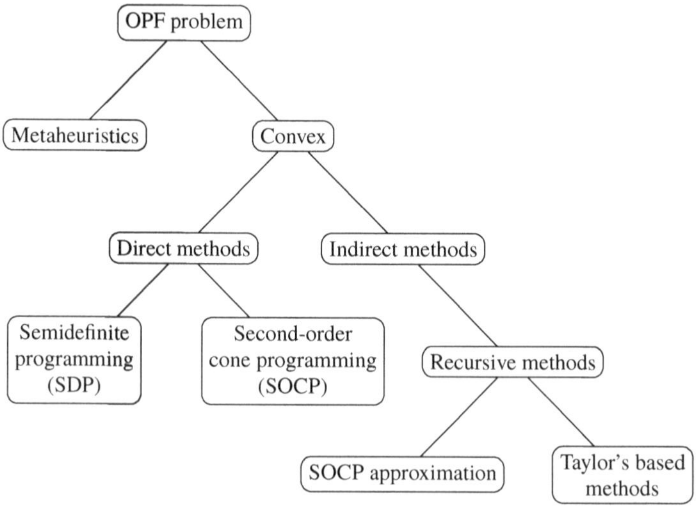


**Fig. 2.3** Possible solution methodologies to deal with the OPF problem in monopolar DC networks

programming [8] and semidefinite programming [17] approaches transform the power flow equations via semidefinite matrices or hyperbolic conic approximations. In the case of indirect methods, linear approximations based on Taylor's series expansions or hyperbolic relaxations are implemented to obtain convex equivalent approximations, which are iteratively solved to minimize the error introduced by linearization point [18].

It is important to mention that in this research, we focus on the application of convex optimization approaches instead of metaheuristic algorithms to solve the OPF problem in monopolar DC networks since convex approaches are mathematically supported; in addition, under well-defined conditions, the optimal global finding is insurable for OPF problems with convex-based approaches. Note that this is not the case for metaheuristic algorithms, which rely on random-based search procedures and therefore cannot guarantee global optimality from a mathematical standpoint [19]. Heuristic algorithms have also been widely criticized by the scientific community due to their lack of rigor [20].

## References

1. Gan, L., & Low, S. H. (2014). Optimal power flow in direct current networks. *IEEE Transactions on Power Systems, 29*(6), 2892–2904.
2. Garces, A. (2017). Uniqueness of the power flow solutions in low voltage direct current grids. *Electric Power Systems Research, 151*, 149–153.

3. Garcés, A., & Montoya, O.-D. (2019). A potential function for the power flow in dc microgrids: An analysis of the uniqueness and existence of the solution and convergence of the algorithms. *Journal of Control, Automation and Electrical Systems, 30*(5), 794–801.
4. Chua, L., Desoer, C., & Ernest, K. U. H. (1987). *Linear and nonlinear circuits*. McGraw-Hill Book Company.
5. Montoya, O. D., Gil-González, W., & Garces, A. (2020). Numerical methods for power flow analysis in DC networks: State of the art, methods and challenges. *International Journal of Electrical Power & Energy Systems, 123*, 106299.
6. Sun, H., Guo, Q., Zhang, B., Guo, Y., Li, Z., & Wang, J. (2014). Master–slave-splitting based distributed global power flow method for integrated transmission and distribution analysis. *IEEE Transactions on Smart Grid, 6*(3), 1484–1492.
7. Grisales-Noreña, L. F., Montoya, O. D., Gil-González, W. J., Perea-Moreno, A.-J., & Perea-Moreno, M.-A. (2020). A comparative study on power flow methods for direct-current networks considering processing time and numerical convergence errors. *Electronics, 9*(12), 2062.
8. Li, J., Liu, F., Wang, Z., Low, S. H., & Mei, S. (2018). Optimal power flow in stand-alone DC microgrids. *IEEE Transactions on Power Systems, 33*(5), 5496–5506.
9. Chen, Y., Qi, H., Li, H., Xu, H., Yang, Q., & Chen, Q. (2022). Convex-optimization-based power-flow calculation method for offshore wind systems. *Energies, 15*(20), 7717.
10. Farivar, M., & Low, S. H. (2013). Branch flow model: Relaxations and convexification—part I. *IEEE Transactions on Power Systems, 28*(3), 2554–2564.
11. Grisales-Noreña, L. F., Rosales-Muñoz, A. A., Cortés-Caicedo, B., Montoya, O. D., & Andrade, F. (2022). Optimal operation of PV sources in DC grids for improving technical, economical, and environmental conditions by using vortex search algorithm and a matrix hourly power flow. *Mathematics, 11*(1), 93.
12. Yuan, Z., & Hesamzadeh, M. R. (2018). Second-order cone AC optimal power flow: Convex relaxations and feasible solutions. *Journal of Modern Power Systems and Clean Energy, 7*(2), 268–280.
13. Garces, A. (2016). A linear three-phase load flow for power distribution systems. *IEEE Transactions on Power Systems, 31*(1), 827–828.
14. Grisales-Noreña, L. F., Garzón-Rivera, O. D., Ocampo-Toro, J. A., Ramos-Paja, C. A., & Rodriguez-Cabal, M. A. (2020). Metaheuristic optimization methods for optimal power flow analysis in DC distribution networks. *Transactions on Energy Systems and Engineering Applications, 1*(1), 13–31.
15. Garces, A. (2018). On the convergence of Newton's method in power flow studies for DC Microgrids. *IEEE Transactions on Power Systems, 33*(5), 5770–5777.
16. Montoya, O. D., Garrido, V. M., Gil-Gonzalez, W., & Grisales-Norena, L. F. (2019). Power flow analysis in DC grids: Two alternative numerical methods. *IEEE Transactions on Circuits and Systems II: Express Briefs, 66*(11), 1865–1869.
17. Garces, A., Montoya, O. D., & Torres, R. (2016). Optimal power flow in multiterminal HVDC systems considering DC/DC converters. In *2016 IEEE 25th international symposium on industrial electronics (ISIE)*. IEEE.
18. Ocampo-Toro, J. A., Montoya, O. D., & Grisales-Noreña, L. F. (2022). Recursive convex approximations for optimal power flow solution in direct current networks. *International Journal of Electrical and Computer Engineering (IJECE) 12*(6), 5674.
19. Sörensen, K. (2015). Metaheuristics—the metaphor exposed. *International Transactions in Operational Research, 22*(1), 3–18.
20. Camacho-Villalón, C. L., Dorigo, M., & Stützle, T. (2022). An analysis of why cuckoo search does not bring any novel ideas to optimization. *Computers & Operations Research, 142*, 105747.

# Chapter 3
# Multiperiod OPF Solution Using Convex Reformulation via Product Approximation

**Abstract** This chapter presents a recursive optimization methodology for solving the optimal power flow (OPF) problem in monopolar DC networks using a novel product-based approximation. By leveraging the multivariable Taylor series expansion, the nonlinear power balance equations in the OPF formulation are transformed into linearized expressions around a predefined operating point. This transformation simplifies the complex, non-convex problem into a convex quadratic programming model suitable for efficient numerical computation. The chapter introduces the linear approximation of the product function, highlights its application in recursive OPF modeling, and provides a detailed Julia-based implementation. Numerical validations are performed on a benchmark test system, including a six-bus MT-HVDC network, to demonstrate the effectiveness of the proposed approach. Key results show that the recursive methodology achieves high accuracy with minimal error compared to exact OPF solutions while maintaining computational efficiency. The iterative process systematically refines the solution, ensuring convergence to an optimal result. The product-based approximation framework is particularly beneficial for modern power systems with dispersed generation and renewable energy integration, as it ensures numerical stability and reduces computational complexity. This chapter concludes with a discussion on the practical implications of the proposed methodology and its potential applications in real-world scenarios, making it a valuable tool for large-scale energy systems analysis and optimization.

**Keywords** Multiperiod optimal power flow · Convex reformulation · Product approximation · Recursive optimization · DC power systems

## 3.1 Multivariable Taylor's Series Expansion

The multivariable linear approximation of a nonlinear, smooth, and differentiable function is essential for deriving an equivalent recursive convex approximation for the optimal power flow (OPF) problem in monopolar DC networks [1]. This linearization is achieved through the multivariable Taylor series expansion, which provides a systematic approach for approximating nonlinear functions in the vicinity

© The Author(s), under exclusive license to Springer Nature Switzerland AG 2026 
O. D. Montoya Giraldo et al., *Optimal Power Flow Methods for DC Networks*, Power Systems, https://doi.org/10.1007/978-3-032-05734-1_3

of a predefined operating point [2]. The linearized form ensures computational tractability and facilitates the use of convex optimization techniques, which are critical for solving large-scale OPF problems efficiently [3].

Consider a two-dimensional nonlinear function $z = f(x, y)$, which can be approximated around the operating point $(x_0, y_0)$ using the first-order terms of the Taylor series expansion, as defined in the equation below:

$$f(x, y) = f(x_0, y_0) + \frac{\partial f(x_0, y_0)}{\partial x}(x - x_0) + \frac{\partial f(x_0, y_0)}{\partial y}(y - y_0) + \text{H.O.T}, \tag{3.1}$$

where H.O.T refers to the higher-order terms of the Taylor series expansion. These higher-order terms are typically neglected to simplify the nonlinear function into a linear approximation. Consequently, Eq. (3.1) represents a linearized (i.e., planar) approximation of the nonlinear function $z = f(x, y)$ in the neighborhood of the point $(x_0, y_0)$.

The linear approximation obtained through this approach is instrumental in OPF formulations, as it transforms inherently nonlinear relationships—such as those involving voltage and power interactions—into linear equations [4]. This linearization not only reduces computational complexity but also ensures that the problem can be solved using efficient convex optimization methods. Furthermore, the recursive application of this linear approximation allows for iterative refinements of the solution, minimizing the approximation error and ensuring that the solution converges to the true optimal value [5]. Such approximations are particularly valuable in electrical networks, where voltage and power variables are often close to their nominal operating points. For instance, in distribution networks with per-unit (p.u.) system representation, voltage magnitudes are typically maintained near 1.0 p.u.

Considering that the nonlinearities in the OPF formulation are associated with the power balance constraint, see Eq. (2.18), in the right-hand side, which corresponds to the product of the voltage variables between each pair of nodes per period; here, we illustrate the linear approximation considering the following function [6]:

$$f(x, y) = xy, \tag{3.2}$$

which can be linearized around $(x_0, y_0) = (1, 1)$ (this operating point is selected since for distribution networks when per-unit representation is used, the voltage variables are near to the unity value, i.e., around 1.0 p.u.) as presented in (3.3).

$$g(x, y) = y_0 x + x_0 y - x_0 y_0 = x + y - 1. \tag{3.3}$$

In addition, the error between the exact function $f(x, y)$ and its linear approximation $g(x, y)$ is defined as $e(x, y) = 100|f(x, y) - g(x, y)|$.

The exact, linear, and error functions regarding the product between two variables are depicted in Fig. 3.1.

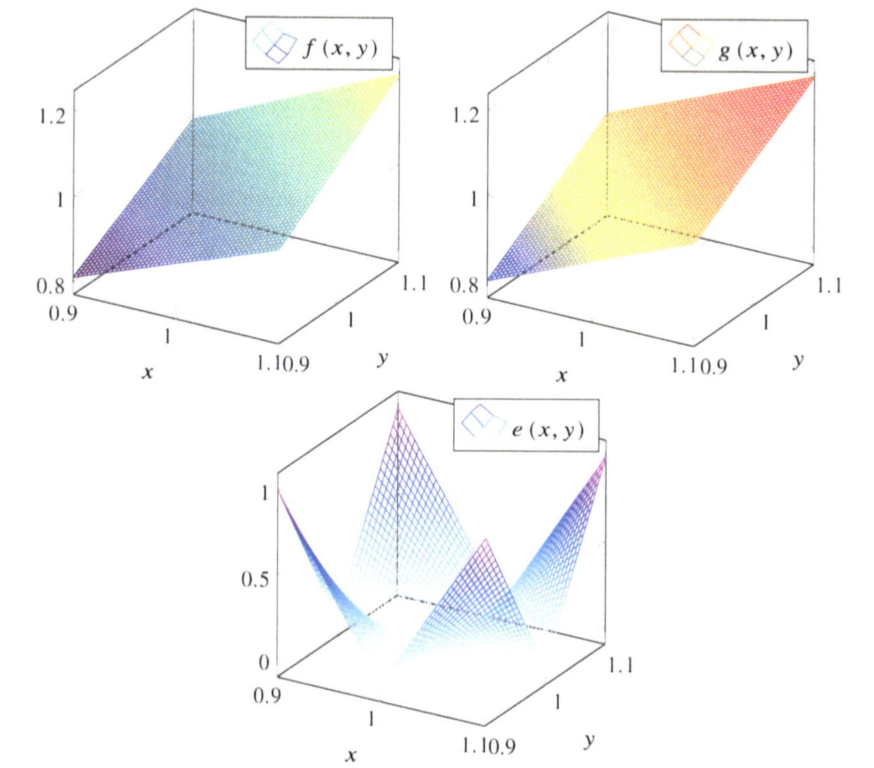


**Fig. 3.1** Three-dimensional representation of the exact, linear, and error functions regarding the product among two variables

From Fig. 3.1, the following observations can be made:

i. The three-dimensional representation of the nonlinear function $f(x, y)$ and its linear approximation $g(x, y)$ around the linearizing point $(1, 1)$ demonstrates a high degree of similarity within the defined bounds of 0.90–1.10. These bounds are typical for voltage regulation limits in electrical networks and effectively illustrate the behavior of both functions in the region of interest.
ii. The error function $e(x, y)$, which quantifies the discrepancy between the nonlinear function $f(x, y)$ and its linear approximation $g(x, y)$, reveals that the maximum error between the two functions is approximately 1% at the edges of the plotting area. This occurs when the variables $x$ and $y$ take values at the boundary limits of 0.90 and 1.10, representing the extremes of the voltage regulation range.
iii. A key feature of the nonlinear function $f(x, y)$ and its linear approximation $g(x, y)$ is that at the linearizing point $(1, 1)$, the error between the two functions is exactly zero. This indicates that under these operating conditions,

the nonlinear function and its linear approximation are numerically equivalent, ensuring accuracy at the selected operating point.

The primary advantage of using the linear approximation $g(x, y)$ to represent the behavior of the nonlinear function $f(x, y)$ around the operating point $(x_0, y_0)$ lies in its iterative refinement capability. When the final values, $(x_f, y_f)$, deviate from the initial linearizing point, the approximation can be progressively updated. This iterative process minimizes or eliminates the approximation error, as represented by the error function $e(x, y)$, ensuring improved accuracy [6]. In the context of the OPF problem, this refinement is efficiently achieved through sequential quadratic programming, which systematically reduces the mismatch between the linear approximation and the original nonlinear function, ultimately enhancing the precision and robustness of the solution.

## 3.2 Recursive OPF Modeling via the Product's Approximation

For the implementation of the recursive OPF solution using the product's approximation, we define the variables $x$ and $y$ in Eq. (3.2) as $x = v_{k,h}$ and $y = v_{m,h}$, where $v_{k,h}$ and $v_{m,h}$ represent the voltages at nodes $k$ and $m$ during time period $h$, respectively. Consequently, the linearizing point at iteration $t$ is expressed as $\left(v_{k,h}^t, v_{m,h}^t\right)$, which corresponds to the voltage values obtained from the previous iteration.

By substituting these definitions into the linearization equation provided in (3.3), the nonlinear product of voltages is approximated linearly around the linearizing point. This transformation is particularly important in the context of the multiperiod OPF model, defined in (2.28) (see Chap. 2), as it reformulates the original nonconvex problem into a convex approximated structure, as presented in Eq. (3.4).

This convex reformulation introduces several advantages, including computational tractability and compatibility with state-of-the-art convex optimization solvers. By leveraging the linearized product approximation, the recursive OPF approach ensures that the optimization model can be solved iteratively, progressively refining the solution to improve accuracy [6].

Each iteration updates the voltage values $v_{k,h}^t$ and $v_{m,h}^t$ based on the solution of the convex model at the preceding step. This iterative process systematically reduces the approximation error introduced by the linearization, ensuring convergence toward an accurate solution for the original nonlinear OPF problem [7]. Furthermore, the use of convex approximations guarantees global optimality at each iteration, making the methodology both robust and efficient for large-scale power systems.

This approach is particularly effective in multiperiod scenarios, where dynamic changes in operating conditions, such as varying load demands or renewable generation outputs, require frequent updates to the voltage and power profiles. By iteratively refining the solution, the recursive OPF framework via the product's approximation enables precise modeling of real-world operational conditions while maintaining computational efficiency.

**Product's Multiperiod OPF Approximation**

$$\text{O.F.:} \quad \min\{E_{\text{loss}}\} = \sum_{k\in\mathbf{N}}\sum_{m\in\mathbf{N}}\sum_{h\in\mathbf{H}} v_{k,h}^{t+1} G_{k,m} v_{m,h}^{t+1} \Delta_h \tag{3.4}$$

st:

$$\begin{bmatrix} p_{k,h}^{s} - p_{k,h}^{dg} \\ -p_{k,h}^{d} \end{bmatrix} = \sum_{m\in\mathbf{N}} G_{km} \begin{pmatrix} v_{k,h}^{t} v_{m,h}^{t+1} + v_{k,h}^{t+1} v_{m,h}^{t} \\ -v_{k,h}^{t} v_{m,h}^{t} \end{pmatrix}, \qquad \{\forall k \in \mathbf{N},\ \forall h \in \mathbf{H}\}$$

$$r_{km} i_{km,h} = v_{k,h}^{t+1} - v_{m,h}^{t+1}, \qquad \{\forall km \in \mathbf{L},\ \forall h \in \mathbf{H}\}$$

$$\left| i_{km,h} \right| \leq i_{km}^{\max}, \qquad \{\forall km \in \mathbf{L},\ \forall h \in \mathbf{H}\}$$

$$p_{k,h}^{s,\min} \leq p_{k,h}^{s} \leq p_{k,h}^{s,\max}, \qquad \{\forall k \in \mathbf{N},\ \forall h \in \mathbf{H}\}$$

$$p_{k,h}^{dg,\min} \leq p_{k,h}^{dg} \leq p_{k,h}^{dg,\max}, \qquad \{\forall k \in \mathbf{N},\ \forall h \in \mathbf{H}\}$$

$$v_{k}^{\min} \leq v_{k,h}^{t+1} \leq v_{k}^{\max}, \qquad \{\forall k \in \mathbf{N},\ \forall h \in \mathbf{H}\}$$

$$v_{k,h}^{t+1} = v_{k,h}^{t} = v_{\text{nom}}, \qquad \{\forall k \in \mathbf{S},\ \forall h \in \mathbf{H}\}.$$

To eliminate the error introduced by Taylor's linearization in the OPF model (3.4), a recursive solution methodology is implemented, by iteratively updating the voltage values $v_{k,h}^{t}$ until they reach the desired convergence between two continuous iterations, i.e.,

$$\max_{k,h} \left\{ \left| v_{k,h}^{t+1} \,||\, - \,||\, v_{k,h}^{t} \right| \right\} \leq \varepsilon. \tag{3.5}$$

## 3.3 Numerical Validations

To validate the effectiveness of the recursive OPF approach via the product's approximation in (3.4), the exact multiperiod OPF model in (2.28) is implemented using the Julia programming environment with the help of its optimization packages `JuMP` and `Ipopt` [8].

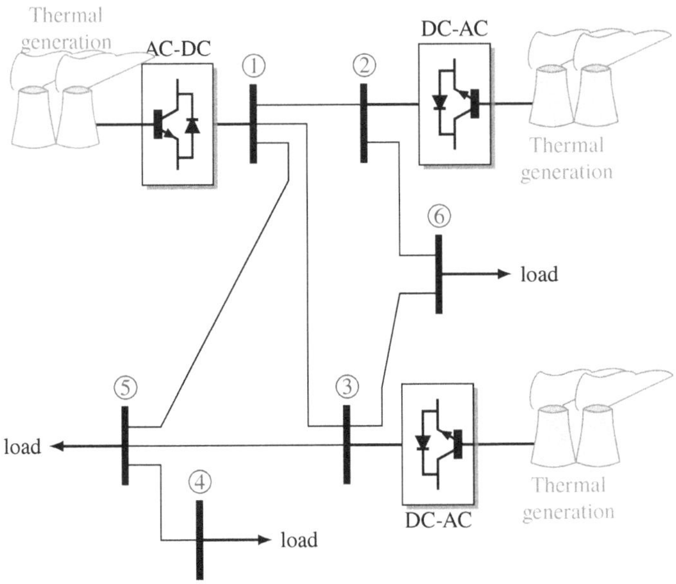


**Fig. 3.2** Electrical configuration of the six-bus system

### 3.3.1 *Six-Bus System*

The six-bus system is a multiterminal high-voltage DC system composed of six nodes, seven lines, and three thermal power plants operated with a monopolar configuration using a voltage level of 400 kV [9]. The electrical configuration of this test system is depicted in Fig. 3.2.

The electrical parametrization of the six-bus test system is listed in Table 3.1, which includes the transmission line resistances, minimum and maximum power bounds for thermal plants, and the peak load consumptions.

Note that for all the transmission lines in this test system, a maximum thermal bound of about 4.75 kA is assigned.

### 3.3.2 *Exact OPF Solution*

The implementation of the exact optimization model for the OPF problem (model (2.28) in Chap. 2) is in the Julia programming considering a single-period analysis for simplicity.

This implementation initializes with the usage of the `Conductance(Lines,L,N)` which allows reaching the value of the conductance system matrix for the test

**Table 3.1** Electrical parameters of the six-bus system

| *Line parameters* | | | | | |
|---|---|---|---|---|---|
| (Line #) From | To | $R_{km}$ (Ω) | From | To | $R_{km}$ (Ω) |
| (Line 1) 1 | 5 | 5.70 | (Line 5) 3 | 6 | 4.75 |
| (Line 2) 5 | 3 | 2.28 | (Line 6) 1 | 2 | 1.90 |
| (Line 3) 5 | 4 | 1.71 | (Line 7) 2 | 6 | 1.90 |
| (Line 4) 1 | 3 | 2.28 | – | – | – |
| *Power generation bounds* | | | | | |
| Node | $p_k^{s,\min}$ (MW) | $p_k^{s,\max}$ (MW) | Node | $p_k^{s,\min}$ (MW) | $p_k^{s,\max}$ (MW) |
| 1 | 50 | 1500 | 2 | 100 | 2000 |
| 3 | 140 | 1800 | – | – | – |
| *Load consumptions* | | | | | |
| Node | $p_k^d$ (MW) | Node | $p_k^d$ (MW) | Node | $p_k^d$ (MW) |
| 4 | 1500 | 5 | 1250 | 6 | 950 |

system under analysis. This implementation is presented below for the six-bus system:

```
import LinearAlgebra
import DataFrames
function Conductance(Lines,L,N)
    Gbus = zeros(N,N);
    for l = 1:L
        k = Lines.k[l,1];  m = Lines.m[l,1];
        rkm = Lines.Rkm[l,1];
        Gbus[k,m] = -1/rkm; Gbus[m,k] = -1/rkm;
        Gbus[k,k] = Gbus[k,k] + 1/rkm;
        Gbus[m,m] = Gbus[m,m] + 1/rkm;
    end
    return Gbus
end
Vb = 400; Pb = 1000; Rb = Vb^2/Pb; Ib = Pb/Vb;
vmin = 0.90; vmax = 1.10;
Lines =  DataFrames.DataFrame([
    (1, 5, 5.70, 4.75),(5, 3, 2.28, 4.75),(5, 4, 1.71, 4.75),
    (1, 3, 2.28, 4.75),(3, 6, 4.75, 4.75),(1, 2, 1.90, 4.75),
    (2, 6, 1.90, 4.75)]);
DataFrames.rename!(Lines, [:k, :m, :Rkm, :Ikm])
Nodes = DataFrames.DataFrame([
    (1, 1, 0,    50,  1500),(2, 1, 0,     100, 2000),
    (3, 1, 0,    140, 1800),(4, 1, 1500, 0,    0),
    (5, 1, 1250, 0,    0),(6, 1, 950,  0,    0),]);
DataFrames.rename!(Nodes, [:k, :Vk0, :Pk, :Pgmin, :Pgmax])
Lines.Rkm= Lines.Rkm/Rb;  Lines.Ikm = Lines.Ikm/Ib;
Nodes.Pk = Nodes.Pk/Pb; Nodes.Pgmin = Nodes.Pgmin/Pb;
Nodes.Pgmax = Nodes.Pgmax/Pb;
N = size(Nodes,1); L = size(Lines,1);
Gbus = Conductance(Lines,L,N);
```

Once the information of the test system was uploaded and the $G_{\text{bus}}$ matrix is obtained, then the nonlinear solver Ipopt using the JuMP optimization environment is used to implement the OPF model (2.28) as presented below:

```
using JuMP, Ipopt
function ExactOPFSolver(Gbus,Nodes,Lines,N,L,vmin,vmax)
    elements = 1:N;  branches = 1:L;
    Pd = Nodes.Pk; Pgmin = Nodes.Pgmin;
    Pgmax = Nodes.Pgmax; Ilmax = Lines.Ikm;
    OPF = Model(Ipopt.Optimizer)
    @variable(OPF,pg[i in elements])
    @variable(OPF,v[i in elements])
    @variable(OPF,il[l in branches])
    @objective(OPF,Min, sum(v[i]*sum(Gbus[i,j]*v[j]
    for j in elements) for i in elements))
    for i = 1:N
        @constraint(OPF,pg[i] >= Pgmin[i])
        @constraint(OPF,pg[i] <= Pgmax[i])
        @constraint(OPF,v[i] <= vmax)
        @constraint(OPF,v[i] >= vmin)
        @NLconstraint(OPF, pg[i] - Pd[i] ==
        v[i]*sum(Gbus[i,j]*v[j] for j in elements))
    end
    @constraint(OPF,v[2] == 1);
    for l = 1:L
        k = Lines.k[l,1]; m = Lines.m[l,1]; rkm = Lines.Rkm[l,1];
        @constraint(OPF,il[l] == (v[k] - v[m])/rkm)
        @NLconstraint(OPF,-Ilmax[l] <= il[l] <= Ilmax[l])
    end
    JuMP.optimize!(OPF)
    pg = [value(pg[i]) for i in elements];
    vol = [value(v[i]) for i in elements];
    il = [value(il[j]) for j in branches];
    fobjval = @show objective_value(OPF)*Pb;
    return vol, il, pg, fobjval;
end
vol,il,pg,fobjval = ExactOPFSolver(Gbus,Nodes,Lines,N,L,vmin,vmax);
Report1 = DataFrames.DataFrame(;
Bus = 1:N, Voltage = vol*Vb, Pgen =pg*Pb);
Report2 = DataFrames.DataFrame(;
Lines = 1:L, From = Lines.k, To = Lines.m,
Current = il*Ib, Loss = Pb*Lines.Rkm.*(il).^2,);
```

Observe that after implementing the exact OPF problem for the six-bus system (i.e., after executing both previous scripts), the numerical results are listed in Table 3.2.

Numerical results in Table 3.2 show that (i) effectively, the output voltage in the slack source is maintained in the reference value, (ii) all the power generators provide power between their minimum and maximum values, (iii) the expected power loss for the six-bus system is about 125.8467 MW, i.e., 3.29% of the total power generation, (iv) the voltage regulation is about 5.20%, which occurs at node 4 (see Fig. 3.2), and it is justifiable due to the radial structure of the MT-HVDC

**Table 3.2** Voltage, power, and current outputs for the six-bus MT-HVDC system

| Node | $v_k$ (kV) | $p_k^s$ (MW) | $p_{loss}$ (MW) | Line | $i_{km}$ (kA) |
|---|---|---|---|---|---|
| 1 | 399.8970 | 1500 | | 1 | 2.4443 |
| 2 | 400.0000 | 900.2383 | | 2 | 4.7500 |
| 3 | 396.7943 | 1425.6085 | | 3 | 3.9557 |
| 4 | 379.2000 | – | 125.8467 | 4 | 1.3608 |
| 5 | 385.9643 | – | | 5 | 0.2037 |
| 6 | 395.8269 | – | | 6 | 0.0542 |
| – | – | – | | 7 | 2.1964 |

system at this node, and (v) transmission line number 2 that connects nodes 3 and 5 presents a loadability of 100%, while transmission line 6 (between nodes 1 and 2) only presents a loadability of 1.14%.

### 3.3.3 Recursive OPF Solution via the Product's Approximation

Here, the implementation of the recursive OPF approximation is presented in model (3.4) is presented. In this validation, we start after the implementation of the test feeder information and the conductance calculation. Next, the OPF model in (3.4) is implemented.

```
using JuMP, Ipopt
function QPM(Gbus,Nodes,Lines,vt,N,L,vmin,vmax)
    elements = 1:N;  branches = 1:L;
    Pd = Nodes.Pk; Pgmin = Nodes.Pgmin;
    Pgmax = Nodes.Pgmax; Ilmax = Lines.Ikm;
    ROPF = Model(Ipopt.Optimizer)
    @variable(ROPF,pg[i in elements])
    @variable(ROPF,v[i in elements])
    @variable(ROPF,il[l in branches])
    @objective(ROPF,Min, sum(v[i]*sum(Gbus[i,j]*v[j]
    for j in elements) for i in elements))
    for i = 1:N
        @constraint(ROPF,pg[i] >= Pgmin[i])
        @constraint(ROPF,pg[i] <= Pgmax[i])
        @constraint(ROPF,v[i] <= vmax)
        @constraint(ROPF,v[i] >= vmin)
        @constraint(ROPF, pg[i] - Pd[i] ==
        sum(Gbus[i,j]*(vt[i]*v[j] + v[i]*vt[j] - vt[i]*vt[j])
        for j in elements))
    end
    @constraint(ROPF,v[2] == 1);
    for l = 1:L
        k = Lines.k[l,1]; m = Lines.m[l,1]; rkm = Lines.Rkm[l,1];
        @constraint(ROPF,il[l] == (v[k] - v[m])/rkm)
        @NLconstraint(ROPF,-Ilmax[l] <= il[l] <= Ilmax[l])
```

```
    end
    JuMP.optimize!(ROPF)
    pg = [value(pg[i]) for i in elements];
    vol = [value(v[i]) for i in elements];
    il = [value(il[j]) for j in branches];
    fobjval = @show objective_value(ROPF)*Pb;
    return vol, il, pg, fobjval;
end
```

Note that the Julia function `QPM(Gbus,Nodes,Lines,vt,N,L,vmin,vmax)` corresponds to the implementation of the quadratic programming model defined in (3.4), where the input `vt` means the voltage profile at iteration $t$, i.e., $v_k^t$, $\forall k \in \mathbf{N}$, which will permit to update the voltage values at the next iteration.

The recursive solution of the OPF problem using the product's approximation is presented below:

```
epsilon = 1e-10; itermax = 100;
function RPA(Gbus,Nodes,Lines,epsilon,itermax)
    vt = ones(N,1);
    for t = 1:itermax
        global vol,il, pg, objval = QPM(Gbus,Nodes,Lines,
                                         vt,N,L,vmin,vmax)
        if maximum(abs.(abs.(vol) - abs.(vt)))< epsilon
            print("Number of iterations = ",t)
            break
        else
            vt = vol
        end
    end
    return vol,il, pg, objval;
end
vol,il,pg,fobjval = RPA(Gbus,Nodes,Lines,epsilon,itermax)
Report1 = DataFrames.DataFrame(;
Bus = 1:N, Voltage = vol*Vb, Pgen =pg*Pb);
Report2 = DataFrames.DataFrame(;
Lines = 1:L, From = Lines.k, To = Lines.m,
Current = il*Ib, Loss = Pb*Lines.Rkm.*(il).^2,);
```

After executing the necessary scripts presented in this subsection, the following results are obtained:

i. The required number of iterations to reach the desired convergence, i.e., $\varepsilon = 1\times^{-10}$, was 6.
ii. The voltage, power generation, power losses, and current flows obtained with the recursive OPF approach via the products approximation present the same numerical performance presented for the exact OPF solution reported in Table 3.2.

## 3.4 Conclusions

In this chapter, a recursive optimization methodology for solving the OPF problem in monopolar DC networks using a product-based approximation was presented. The proposed methodology leveraged the multivariable Taylor series expansion to linearize nonlinear power balance equations, transforming the original non-convex problem into a convex quadratic programming model. The recursive framework systematically updated the voltage profiles at each iteration, progressively reducing the approximation error and ensuring convergence to an accurate solution. Numerical validations on a six-bus MT-HVDC system demonstrated the effectiveness of the approach, achieving a high level of accuracy with minimal computational effort. The results showed that the recursive OPF method closely approximates the exact solution while maintaining numerical stability and computational efficiency. This methodology is particularly valuable for large-scale power systems with dispersed generation and renewable energy integration, offering a practical and robust tool for addressing the challenges posed by modern power system operations.

## References


1. Javadi, M. S., Gouveia, C. S., Carvalho, L. M., & Silva, R. (2021). Optimal power flow solution for distribution networks using quadratically constrained programming and McCormick relaxation technique. In *2021 IEEE International Conference on Environment and Electrical Engineering and 2021 IEEE Industrial and Commercial Power Systems Europe (EEEIC /&CPS Europe)* (pp. 1–6). IEEE.
2. Duistermaat, J. J., & Kolk, J. A. C. (2010). *Taylor Expansion in Several Variables* (pp. 59–63). Birkhäuser Boston.
3. Boyd, S., & Vandenberghe, L. (2004). *Convex optimization*. Cambridge University Press.
4. Montoya, O. D., Garrido, V. M., Gil-Gonzalez, W., & Grisales-Norena, L. F. (2019). Power flow analysis in DC grids: Two alternative numerical methods. *IEEE Transactions on Circuits and Systems II: Express Briefs, 66*(11), 1865–1869.
5. Sadat, S. A., & Sahraei-Ardakani, M. (2021). Customized sequential quadratic programming for solving large-scale AC optimal power flow. In *2021 North American Power Symposium (NAPS)* (pp. 1–6). IEEE.
6. Montoya, O. D., Gil-González, W., & Garces, A. (2020). Numerical methods for power flow analysis in DC networks: State of the art, methods and challenges. *International Journal of Electrical Power & Energy Systems, 123*, 106299.
7. Jiang, Q., & Chiang, H.-D. (2008). A sequential feasible optimal power flow in power systems. In *2008 IEEE Power and Energy Society General Meeting—Conversion and Delivery of Electrical Energy in the 21st Century* (pp. 1–6). IEEE.
8. Lubin, M., Dowson, O., Dias Garcia, J., Huchette, J., Legat, B., & Vielma, J. P. (2023). JuMP 1.0: Recent improvements to a modeling language for mathematical optimization. *Mathematical Programming Computation, 15*, 581–589.
9. Contreras-Sepúlveda, W., Montoya, O. D., & Gil-González, W. (2024). An economic-environmental energy management system design for MT-HVDC networks via a semi-definite programming approximation with robust analysis. *Ain Shams Engineering Journal, 15*(9), 102968.

# Chapter 4
# Multiperiod OPF Solution Through Convex Reformulation via Hyperbolic Approximation

**Abstract** This chapter introduces the hyperbolic power flow approximation for monopolar DC networks, a novel approach designed to address the inherent nonlinearity in power flow equations. The methodology involves applying Taylor's series expansion to the hyperbolic relationship between voltages and powers in the power equilibrium constraint, enabling the transformation of the original non-convex problem into a simplified and computationally tractable form. By leveraging a first-order Taylor series expansion, the power-voltage relation is approximated with sufficient accuracy for initial calculations. To mitigate the approximation errors introduced by the first-order expansion, a recursive solution framework is employed, iteratively refining the results to ensure convergence toward a highly accurate solution. The proposed hyperbolic approximation offers significant advantages, including reduced computational complexity and enhanced numerical stability, making it particularly suitable for large-scale network analysis. The recursive process systematically improves the quality of the solution, reducing the mismatch between the approximated and true power flow equations. To validate the effectiveness and robustness of this approach, a numerical case study is conducted using Julia, a high-performance programming language well suited for mathematical modeling and optimization. The results of the case study highlight the ability of the hyperbolic approximation to achieve accurate and efficient solutions for optimal power flow (OPF) problems in monopolar DC networks, demonstrating its potential for real-world applications in modern power systems.

**Keywords** Hyperbolic approximation · Convex reformulation · Recursive optimization · DC power flow · Multiperiod optimal power flow

## 4.1 Single-Variable Taylor Series Expansion

The Taylor series expansion for a single-variable function allows the approximation of a nonlinear function $y = f(x)$ around a specific operating point $x_0$ by expressing it as a polynomial [1]. The general form of the expansion is given by

© The Author(s), under exclusive license to Springer Nature Switzerland AG 2026 
O. D. Montoya Giraldo et al., *Optimal Power Flow Methods for DC Networks*, Power Systems, https://doi.org/10.1007/978-3-032-05734-1_4

$$f(x) = f(x_0) + \frac{\partial f(x)}{\partial x}\Big|_{x=x_0}(x - x_0) + \text{H.O.T.}, \tag{4.1}$$

where H.O.T. represents higher-order terms. By neglecting the higher-order terms, a linear approximation of the nonlinear function $f(x)$ around $x_0$ is obtained, simplifying the representation and enabling its use in practical applications [2].

In the specific context of multiperiod OPF analysis, the Taylor series approximation is employed to linearize the hyperbolic inverse relationship between voltage and power. The function of interest is $f(x) = 1/x$, where $x$ represents the voltage, and the approximation is centered around an operating point $x_0$. By substituting this function into the Taylor series and retaining only the first-order terms, the linear equivalent $g(x)$ of $f(x)$ around $x_0$ is derived as

$$g(x) = \frac{1}{x_0} - \frac{1}{x_0^2}(x - x_0) = \frac{2}{x_0} - \frac{x}{x_0^2}, \tag{4.2}$$

where the approximation preserves the essential characteristics of the inverse function while simplifying its computation [3].

This linearization is critical in OPF formulations as it transforms the non-convex hyperbolic relation into a linear form, enabling the application of convex optimization techniques [2]. The linearized model facilitates efficient numerical computation and iterative refinement, particularly in scenarios involving multiple periods, dispersed generation, or dynamic operating conditions. Such approximations play a pivotal role in balancing computational tractability and solution accuracy in modern power system analysis [4].

In the context of the OPF analysis in monopolar DC networks, the main idea is to approximate the voltage profile within its operating region, typically around $v_k^0 \approx 1$ p.u.. To illustrate the effect of the proposed linearization, defined by $g(x)$ over $f(x)$, let us consider $x_0 = 1$. In addition, the error between both functions is defined as $e(x) = 100|\frac{1}{x} - 2 + x|$. The behavior of these functions in the area of interest for electrical networks, i.e., $x_{\min} \leq x \leq x_{\max}$, with $x_{\min} = 0.90$ p.u. and $x_{\max} = 1.10$ p.u., is depicted in Fig. 4.1.

The behavior of the exact, linear, and error functions in Fig. 4.1 shows that (i) in the region of interest, the exact function $f(x)$ and the linear approximation $g(x)$ present a very similar tendency, especially when the $x$ variable is near the linearizing point, i.e., $0.96 \leq x \leq 1.04$, and (ii) in this region of operation the maximum error between both functions (the exact and linear approximations) defined by $e(x)$ is less than 0.20%, and the maximum error when the $x$ variable is in its minimum and maximum values, i.e., 0.90 or 1.10, is about 1.11% and 0.91%.

The primary advantage of using the linear approximation $g(x)$ to represent the behavior of the nonlinear function $f(x)$ around the operating point $x_0$ lies in its flexibility and iterative refinement capability. If the final value, $x_f$, deviates from the initial linearization point $x_0$, the operating point can be iteratively updated to reduce or eliminate the approximation error. This error, represented by the function $e(x)$, can be minimized through sequential quadratic programming (SQP)

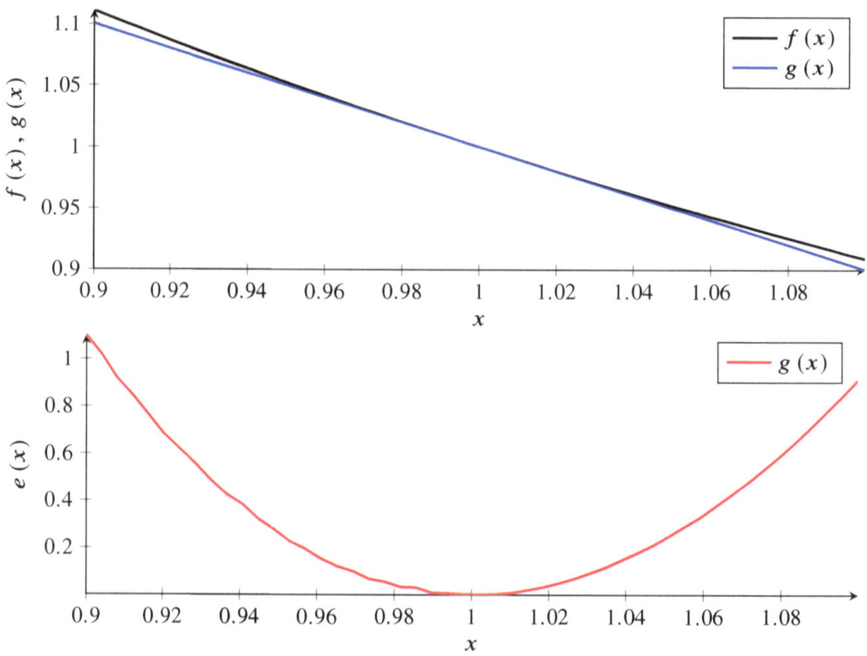


**Fig. 4.1** Bidimensional representation of the exact $f(x)$, linear $g(x)$, and error $e(x)$ functions around $x_0 = 1$

methods, particularly in the context of solving the OPF problem [5]. This iterative approach ensures that the linearized model remains accurate, even as the operating conditions deviate from the initial assumptions, ultimately improving the precision and reliability of the solution [6].

## 4.2 Recursive OPF Modeling via the Hyperbolic Approximation

To implement the recursive OPF solution via the hyperbolic approximation, we define the variable $x$ in Eq. (3.2) as $x = v_{k,h}$, where $v_{k,h}$ represents the voltage at bus $k$ during time period $h$. Based on this definition, the linearization point at iteration $t$ is expressed as $x_0 = v^t_{k,h}$, representing the voltage value from the previous iteration. This iterative linearization approach ensures that the model progressively refines its approximation as the solution evolves over iterations.

By applying the linearization formula from Eq. (4.2) to the hyperbolic relation, the nonlinear term in the power-voltage relationship is approximated with a first-order Taylor series expansion. This transformation allows the original multiperiod OPF model, defined in Eq. (2.28) (refer to Chap. 2), to be reformulated into a convex structure, as shown in Eq. (4.3). The convex approximation is particularly advantageous because it ensures global optimality, computational tractability, and compatibility with efficient conic optimization solvers.

This recursive approach systematically reduces the error introduced by the linearization process at each iteration, as the voltage values are updated and the linearization point shifts closer to the true solution. The process is repeated until convergence is achieved, resulting in an accurate and computationally efficient solution to the OPF problem. This methodology highlights the practical utility of hyperbolic approximations in transforming complex, nonlinear OPF models into convex formulations suitable for large-scale power systems.

**Hyperbolic Multiperiod OPF Approximation**

$$\text{O.F.: } \min\left\{E_{\text{loss}}\right\} = \sum_{k\in\mathbf{N}}\sum_{m\in\mathbf{N}}\sum_{h\in\mathbf{H}} v_{k,h}^{t+1} G_{k,m} v_{m,h}^{t+1} \Delta_h \tag{4.3}$$

st:

$$\frac{p_{k,h}^{s}}{v_{k,h}^{t}} - \frac{p_{k,h}^{dg}}{v_{k,h}^{t}} - \left(\frac{2}{v_{k,h}^{t}} - \frac{v_{k,h}^{t+1}}{\left(v_{k,h}^{t}\right)^{2}}\right) p_{k,h}^{d} = \sum_{m\in\mathbf{N}} G_{km} v_{m,h}^{t+1}, \qquad \{\forall k \in \mathbf{N},\ \forall h \in \mathbf{H}\}$$

$$r_{km} i_{km,h} = v_{k,h}^{t+1} - v_{m,h}^{t+1}, \qquad \{\forall km \in \mathbf{L},\ \forall h \in \mathbf{H}\}$$

$$\left|i_{km,h}\right| \leq i_{km}^{\max}, \qquad \{\forall km \in \mathbf{L},\ \forall h \in \mathbf{H}\}$$

$$p_{k,h}^{s,\min} \leq p_{k,h}^{s} \leq p_{k,h}^{s,\max}, \qquad \{\forall k \in \mathbf{N},\ \forall h \in \mathbf{H}\}$$

$$p_{k,h}^{dg,\min} \leq p_{k,h}^{dg} \leq p_{k,h}^{dg,\max}, \qquad \{\forall k \in \mathbf{N},\ \forall h \in \mathbf{H}\}$$

$$v_{k}^{\min} \leq v_{k,h}^{t+1} \leq v_{k}^{\max}, \qquad \{\forall k \in \mathbf{N},\ \forall h \in \mathbf{H}\}$$

$$v_{k,h}^{t+1} = v_{k,h}^{t} = v_{\text{nom}}, \qquad \{\forall k \in \mathbf{S},\ \forall h \in \mathbf{H}\}.$$

In the optimization model (4.3), the following assumption regarding the hyperbolic relations between the power generation outputs and the voltage profiles was considered:

$$h\left(x, y\right) = \frac{y}{x}, \tag{4.4}$$

where $y$ represents an unknown variable (such as power generation in the context of the OPF analysis), and the term $\frac{1}{x}$ is approximated around a known operating point $x_0$ to achieve a linear relationship between the two variables. This approximation leads to the reformulation:

$$h\left(x,y\right) \approx \frac{y}{x_0}. \tag{4.5}$$

This linearization simplifies the mathematical structure of the hyperbolic relation, making it more tractable for optimization algorithms while preserving the essential characteristics of the original nonlinear equation.

To address and eliminate the error introduced by the Taylor series linearization in the OPF model (4.3), a recursive solution methodology is employed. This iterative approach involves updating the voltage values $v_{k,h}^t$ at each iteration $t$, refining the linearized approximation until the desired level of convergence is achieved. The convergence criterion is defined as the maximum deviation between the voltage magnitudes across two consecutive iterations being less than a predefined tolerance $\varepsilon$, as expressed by

$$\max_{k,h}\left\{\left|v_{k,h}^{t+1}\right| - \left|v_{k,h}^{t}\right|\right\} \leq \varepsilon. \tag{4.6}$$

This recursive process ensures that the solution gradually approaches the true optimal solution of the original nonlinear OPF problem. By iteratively refining the voltage estimates, the method systematically reduces the approximation error introduced by the linearization, allowing the solution to align closely with the exact power flow equations. The recursive methodology also takes advantage of modern computational tools to efficiently handle large-scale systems while ensuring precision.

The use of a small convergence tolerance $\varepsilon$ guarantees high accuracy in the final solution without compromising computational efficiency. This balance between precision and efficiency is particularly advantageous for practical applications in real-world power systems, where system dynamics and uncertainties often necessitate iterative refinement to achieve reliable solutions [7]. Furthermore, the recursive hyperbolic approximation provides a robust framework for addressing the nonlinearities inherent in OPF problems, ensuring that the proposed model remains applicable to diverse scenarios, including networks with varying load conditions, dispersed generation, and renewable energy integration.

## 4.3 Numerical Validations

The validation of the hyperbolic OPF formulation in (4.3) is made by comparing its results with the exact OPF model solved in Chap. 3. The numerical implementations are made with the help of the `JuMP` optimization package and the `Ipopt` solver in the Julia programming language [8].

**Fig. 4.2** 11-node MT-HVDC system

### *4.3.1 The 11-Bus System*

The 11-bus system under consideration is a multiterminal high-voltage direct current (MT-HVDC) network designed to operate at a nominal voltage of 380 kV. This network consists of 11 nodes interconnected by 17 high-capacity transmission lines, forming a robust and flexible grid structure [9]. The system includes three thermal power plants, which provide dispatchable generation to maintain grid stability, and two renewable generators based on photovoltaic (PV) technology, contributing to the integration of clean energy sources. Additionally, six nodes in the network are configured as constant power loads, representing typical industrial and residential consumption patterns.

The electrical configuration of this MT-HVDC system, shown in Fig. 4.2, highlights the strategic placement of generation and load nodes, as well as the interconnection of the transmission lines. This configuration enables the evaluation of advanced control strategies and optimization techniques for power flow, voltage stability, and energy management in MT-HVDC networks [10]. The integration of renewable energy sources and conventional generation technologies within this system makes it an ideal benchmark for assessing the challenges and opportunities associated with modern high-voltage DC grids.

The electrical parametrization of the six-bus test system is listed in Table 4.1, which includes the transmission line resistances, minimum and maximum power bounds for thermal plants, and the peak load consumptions.

Note that for the 11-bus system, the slack bus assigned at bus 3 with a voltage-controlled output of 380 kV, i.e., 1.0 p.u.

**Table 4.1** Parametric information regarding branch resistances and peak load consumptions

| *Line parameters* | | | | | |
|---|---|---|---|---|---|
| (Line #) From | To | $R_{km}$ (Ω) - $I_{km}^{\max}$ (kA) | From | To | $R_{km}$(Ω) - $I_{km}^{\max}$ (kA) |
| (Line 1) 1 | 2 | 3.85 (3.50) | (Line 10) 5 | 9 | 3.34 (3.00) |
| (Line 2) 1 | 4 | 4.22 (3.20) | (Line 11) 5 | 10 | 4.12 (2.80) |
| (Line 3) 1 | 6 | 4.85 (3.10) | (Line 12) 5 | 11 | 3.78 (2.00) |
| (Line 4) 2 | 6 | 2.37 (3.50) | (Line 13) 6 | 7 | 4.65 (3.00) |
| (Line 5) 2 | 7 | 3.25 (3.10) | (Line 14) 6 | 11 | 5.25 (2.60) |
| (Line 6) 3 | 7 | 2.95 (2.90) | (Line 15) 7 | 8 | 3.14 (3.00) |
| (Line 7) 3 | 9 | 4.36 (3.20) | (Line 16) 8 | 9 | 4.55 (1.50) |
| (Line 8) 4 | 6 | 4.02 (3.50) | (Line 17) 8 | 11 | 5.14 (1.60) |
| (Line 9) 4 | 10 | 3.87 (3.60) | – | – | – |
| *Power generation bounds in thermal sources* | | | | | |
| Node | $p_k^{s,\min}$ (MW) | $p_k^{s,\max}$ (MW) | Node | $p_k^{s,\min}$ (MW) | $p_k^{s,\max}$ (MW) |
| 1 | 150 | 1350 | 2 | 300 | 2400 |
| 3 | 250 | 1800 | – | – | – |
| *Power generation bounds in PV plants* | | | | | |
| Node | $p_k^{dg,\min}$ (MW) | $p_k^{dg,\max}$ (MW) | Node | $p_k^{dg,\min}$ (MW) | $p_k^{dg,\max}$ (MW) |
| 4 | 0 | 2500 | 5 | 0 | 2000 |
| *Load consumptions* | | | | | |
| Node | $P$ [MW] | Node | $P$ [MW] | Node | $P$ [MW] |
| 6 | 850 | 7 | 750 | 8 | 950 |
| 9 | 800 | 10 | 650 | 11 | 700 |

### 4.3.2 *Recursive OPF Solution via the Product's Approximation*

This implementation initializes with the usage of the `Conductance(Lines,L,N)` which allows reaching the value of the conductance system matrix for the test system under analysis. This implementation is presented below for the 11-bus system:

```
import LinearAlgebra
import DataFrames
function Conductance(Lines,L,N)
    Gbus = zeros(N,N);
    for l = 1:L
        k = Lines.k[l,1];  m = Lines.m[l,1];
        rkm = Lines.Rkm[l,1];
        Gbus[k,m] = -1/rkm; Gbus[m,k] = -1/rkm;
        Gbus[k,k] = Gbus[k,k] + 1/rkm;
        Gbus[m,m] = Gbus[m,m] + 1/rkm;
    end
    return Gbus
end
Vb = 380; Pb = 1000; Rb = Vb^2/Pb;
Ib = Pb/Vb; vmin = 0.90; vmax = 1.10;
```

```
Lines =  DataFrames.DataFrame([
    (1, 2, 3.85, 3.50),(1, 4,  4.22, 3.20),(1, 6,  4.85, 3.10),
    (2, 6, 2.37, 3.50),(2, 7,  3.25, 3.10),(3, 7,  2.95, 2.90),
    (3, 9, 4.36, 3.20),(4, 6,  4.02, 3.50),(4, 10, 3.87, 3.60),
    (5, 9, 3.34, 3.00),(5, 10, 4.12, 2.80),(5, 11, 3.78, 2.00),
    (6, 7, 4.65, 3.00),(6, 11, 5.25, 2.60),(7, 8,  3.14, 3.00),
    (8, 9, 4.55, 1.50),(8, 11, 5.14, 1.60)]);
    DataFrames.rename!(Lines, [:k, :m, :Rkm, :Ikm])
Nodes = DataFrames.DataFrame([
    (1,  1, 0, 150, 1350),(2,  1, 0,   300, 2400),
    (3,  1, 0, 250, 1800),(4,  1, 0,   0,   2500),
    (5,  1, 0, 0,   2000),(6,  1, 850, 0,   0),
    (7,  1, 750, 0, 0),   (8,  1, 950, 0,   0),
    (9,  1, 800, 0, 0),   (10, 1, 650, 0,   0),
    (11, 1, 700, 0, 0.0)]);
DataFrames.rename!(Nodes, [:k, :Vk0, :Pk, :Pgmin, :Pgmax])
Lines.Rkm= Lines.Rkm/Rb;  Lines.Ikm = Lines.Ikm/Ib;
Nodes.Pk = Nodes.Pk/Pb; Nodes.Pgmin = Nodes.Pgmin/Pb;
Nodes.Pgmax = Nodes.Pgmax/Pb;
N = size(Nodes,1); L = size(Lines,1);
Gbus = Conductance(Lines,L,N);
```

Once the information of the test system was uploaded and the $G_{\text{bus}}$ matrix is obtained, then the nonlinear solver Ipopt using the JuMP optimization environment is used to implement the OPF model (4.3) as presented below:

```
using JuMP, Ipopt
function QPM(Gbus,Nodes,Lines,vt,N,L,vmin,vmax)
    elements = 1:N;  branches = 1:L;
    Pd = Nodes.Pk; Pgmin = Nodes.Pgmin;
    Pgmax = Nodes.Pgmax; Ilmax = Lines.Ikm;
    ROPF = Model(Ipopt.Optimizer)
    @variable(ROPF,pg[i in elements])
    @variable(ROPF,v[i in elements])
    @variable(ROPF,il[l in branches])
    @objective(ROPF,Min, sum(v[i]*sum(Gbus[i,j]*v[j]
    for j in elements) for i in elements))
    for i = 1:N
        @constraint(ROPF,pg[i] >= Pgmin[i])
        @constraint(ROPF,pg[i] <= Pgmax[i])
        @constraint(ROPF,v[i] <= vmax)
        @constraint(ROPF,v[i] >= vmin)
        @NLconstraint(ROPF, pg[i]/vt[i] -
        (2/vt[i] - v[i]/(vt[i])^2)*Pd[i] == sum(Gbus[i,j]*v[j]
        for j in elements))
    end
    @constraint(ROPF,v[3] == 1);
    for l = 1:L
        k = Lines.k[l,1]; m = Lines.m[l,1]; rkm = Lines.Rkm[l,1];
        @constraint(ROPF,il[l] == (v[k] - v[m])/rkm)
        @NLconstraint(ROPF,-Ilmax[l] <= il[l] <= Ilmax[l])
    end
    JuMP.optimize!(ROPF)
    pg = [value(pg[i]) for i in elements];
```

```
    vol = [value(v[i]) for i in elements];
    il = [value(il[j]) for j in branches];
    fobjval = @show objective_value(ROPF)*Pb;
    return vol, il, pg, fobjval;
end
```

Note that the Julia function `QPM(Gbus,Nodes,Lines,vt,N,L,vmin,vmax)` corresponds to the implementation of the quadratic programming model defined in (4.3) (i.e., the hyperbolic OPF approximation), where the input `vt` means the voltage profile at iteration $t$, i.e., $v_k^t$, $\forall k \in \mathbf{N}$, which will permit to update the voltage values at the next iteration.

The recursive solution of the OPF problem using the hyperbolic approximation is presented below:

```
epsilon = 1e-10; itermax = 100;
function RHA(Gbus,Nodes,Lines,epsilon,itermax)
    vt = ones(N,1);
    for t = 1:itermax
        global vol,il, pg, objval = QPM(Gbus,Nodes,Lines,
                                        vt,N,L,vmin,vmax)
        if maximum(abs.(abs.(vol) - abs.(vt)))< epsilon
            print("Number of iterations = ",t)
            break
        else
            vt = vol
        end
    end
    return vol,il, pg, objval;
end
vol,il,pg,fobjval = RHA(Gbus,Nodes,Lines,epsilon,itermax)
Report1 = DataFrames.DataFrame(;
Bus = 1:N, Voltage = vol*Vb, Pgen =pg*Pb);
Report2 = DataFrames.DataFrame(;
Lines = 1:L, From = Lines.k, To = Lines.m,
Current = il*Ib, Loss = Pb*Lines.Rkm.*(il).^2,);
```

After executing the recursive hyperbolic OPF approximation, the results presented in Table 4.2 are obtained.

These numerical results highlight the following key observations:

i. As expected, the voltage profile is fixed at 380 kV at node 3, serving as the reference point for the system. The minimum voltage value in the network occurs at node 8, with a magnitude of approximately 371.3299 kV. This corresponds to a voltage regulation bound of 2.28%, indicating that the system maintains voltage stability within acceptable limits.
ii. The total power losses in the system are approximately 70.4289 MW, reflecting a high transmission efficiency of 98.5236%. This demonstrates the effective performance of the MT-HVDC system in minimizing energy dissipation during power transmission.
iii. The minimum loadability among the transmission lines is observed on line 1, with a utilization of just 0.1828%, indicating a significantly underutilized

**Table 4.2** Voltage, power, and current outputs for the 11-node MT-HVDC system

| Node | $v_k$ (kV) | $p_k^s$ (MW) | $p_{loss}$ (MW) | Line | $i_{km}$ (kA) | Line | $i_{km}$ (kA) |
|---|---|---|---|---|---|---|---|
| Node | $v_k$ (kV) | $p_k^s$ (MW) | $p_{loss}$ (MW) | Line | $i_{km}$ (kA) | Line | $i_{km}$ (kA) |
| 1 | 380.4174 | 281.1379 | | 1 | 0.0064 | 12 | 1.7624 |
| 2 | 380.3928 | 1140.8448 | | 2 | 0.0047 | 13 | 0.2896 |
| 3 | 380.0000 | 1035.6742 | | 3 | 0.7373 | 14 | 0.5851 |
| 4 | 380.4372 | 680.5457 | | 4 | 1.4985 | 15 | 1.3264 |
| 5 | 380.4319 | 1632.2264 | | 5 | 1.5071 | 16 | 0.7573 |
| 6 | 376.8414 | – | 70.4289 | 6 | 1.5272 | 17 | 0.4747 |
| 7 | 375.4949 | – | | 7 | 1.1983 | – | – |
| 8 | 371.3299 | – | | 8 | 0.8945 | – | – |
| 9 | 374.7754 | – | | 9 | 0.8897 | – | – |
| 10 | 376.9934 | – | | 10 | 1.6936 | – | – |
| 11 | 373.7699 | – | | 11 | 0.8345 | – | – |

capacity. In contrast, the maximum loadability is found on line 12, with a utilization of 88.1222%, highlighting its critical role in the power transfer process.

iv. The recursive hyperbolic OPF approach required 41 iterations to achieve the desired convergence, with a tolerance set at $\varepsilon = 1e - 10$. This demonstrates the iterative method's ability to refine the solution to a high degree of accuracy.

## 4.4 Conclusions

The recursive hyperbolic OPF approximation demonstrates its capability to effectively address the nonlinear nature of optimal power flow problems in monopolar DC networks. By iteratively refining the solution, the method ensures high accuracy and convergence to near-optimal results. The voltage profile analysis reveals that the system maintains stability, with a voltage regulation bound of 2.28% and the minimum voltage occurring at node 8, a value well within acceptable operational limits. Furthermore, the system achieves a high transmission efficiency of 98.5236%, with total power losses amounting to 70.4289 MW, emphasizing the method's ability to minimize energy dissipation in large-scale MT-HVDC networks.

The results also highlight disparities in transmission line utilization, with line 1 operating at a minimal loadability of 0.1828%, indicating underutilization, while line 12 approaches its capacity with 88.1222%, emphasizing its critical role in the network. This insight suggests the need for further optimization in power flow distribution to improve line loadability balance. Additionally, the recursive approach successfully converged within 41 iterations, demonstrating its computational efficiency and robustness in solving complex OPF problems with a tight tolerance of $\varepsilon = 1e - 10$.

Overall, the numerical findings validate the practical utility of the hyperbolic approximation method in simplifying nonlinear OPF problems, ensuring computational tractability, and delivering accurate and efficient solutions. These results highlight the potential of this methodology for broader applications in real-world DC power systems, particularly those incorporating renewable energy sources and operating under dynamic conditions.

## References

1. Grisales-Noreña, L. F., Montoya, O. D., Gil-González, W. J., Perea-Moreno, A.-J., & Perea-Moreno, M.-A. (2020). A comparative study on power flow methods for direct-current networks considering processing time and numerical convergence errors. *Electronics, 9*(12), 2062.
2. Garces, A. (2016). A linear three-phase load flow for power distribution systems. *IEEE Transactions on Power Systems, 31*(1), 827–828.
3. Montoya, O. D., Grisales-Noreña, L. F., González-Montoya, D., Ramos-Paja, C. A., & Garces, A. (2018). Linear power flow formulation for low-voltage DC power grids. *Electric Power Systems Research, 163*, 375–381.
4. Altorok, M. , Ukil, A., Nair, N.-K. C., & Rayudu, R. (2024). Review on optimal power flow in a hybrid AC/DC distribution network. *Electric Power Systems Research, 236*, 110956.
5. Sadat, S. A., & Sahraei-Ardakani, M. (2021). Customized sequential quadratic programming for solving large-scale AC optimal power flow. In *2021 North American Power Symposium (NAPS)* (pp. 1–6). IEEE.
6. Hasan, M. S., Chowdhury, M. M.-U.-T., & Kamalasadan, S. (2024). Sequential Quadratic Programming (SQP) based optimal power flow methodologies for electric distribution system with high penetration of DERs. *IEEE Transactions on Industry Applications, 60*(3), 4810–4820.
7. Grisales-Noreña, L. F., Garz6n-Rivera, O. D., Ocampo-Toro, J. A., Ramos-Paja, C. A., & Rodriguez-Cabal, M. A. (2020). Metaheuristic optimization methods for optimal power flow analysis in DC distribution networks. *Transactions on Energy Systems and Engineering Applications, 1*(1), 13–31.
8. Lubin, M. , Dowson, O., Dias Garcia, J., Huchette, J., Legat, B., & Vielma, J. P. (2023). JuMP 1.0: Recent improvements to a modeling language for mathematical optimization. *Mathematical Programming Computation, 15*(3), 581–589.
9. Contreras-Sepúlveda, W., Montoya, O. D., & Gil-González, W. (2024). An economic-environmental energy management system design for MT-HVDC networks via a semi-definite programming approximation with robust analysis. *Ain Shams Engineering Journal, 15*(9), 102968.
10. Garces, A., & Azhmyakov, V. (2020). Application of the nested convex programming to the optimal power flow in MT-HVDC grids. *IFAC-PapersOnLine, 53*(2), 13173–13177.

# Chapter 5
# Multiperiod OPF Solution Through Convex Reformulation via a Recursive Conic Approximation

**Abstract** This chapter introduces a recursive optimal power flow (OPF) formulation tailored for monopolar DC networks. The methodology leverages an innovative conic relaxation approach to accurately model constant power loads. Specifically, these loads are represented using an auxiliary variable that captures the expected demanded current, enabling its relationship with voltages and powers to be expressed through a convex conic formulation. To address the inherent nonlinearity between power and voltage at the generation nodes, an iterative relaxation mechanism is employed. At each iteration $t$, the expected voltage value is updated and used as a reference to compute the next voltage profile by solving an equivalent conic approximation of the OPF model. This recursive process systematically reduces or eliminates the approximation error introduced by the initial voltage assumptions, ensuring convergence to a near-optimal solution. The proposed methodology is particularly advantageous for improving computational efficiency and scalability in the analysis of DC networks, where conventional methods struggle to balance accuracy and tractability. To validate the efficacy and robustness of the proposed approach, a numerical case study is implemented using Julia, a high-performance programming language ideal for mathematical optimization. Results from the case study highlight the model's ability to achieve precise OPF solutions while maintaining computational efficiency, underscoring its potential for real-world applications in DC power systems. This chapter contributes to the growing body of research on conic optimization techniques in power systems, providing a promising framework for addressing the challenges of OPF in DC networks with constant power loads.

**Keywords** Conic optimization · Recursive OPF · Constant power loads · DC power flow · Convex approximation

## 5.1 Second-Order Cone Definition

In convex optimization theory, multiple cone representations play a fundamental role in the formulation and solution of linear, conic, and semidefinite programming

© The Author(s), under exclusive license to Springer Nature Switzerland AG 2026

O. D. Montoya Giraldo et al., *Optimal Power Flow Methods for DC Networks*, Power Systems, https://doi.org/10.1007/978-3-032-05734-1_5

problems [1]. These representations are critical for modeling and solving a variety of engineering and scientific problems efficiently, as they ensure convexity and allow the use of advanced optimization techniques, such as interior-point methods, to achieve globally optimal solutions. Among these representations, the second-order cone (SOC) is particularly significant due to its ability to model a wide range of practical constraints, including quadratic and hyperbolic relationships, in a mathematically tractable manner [2].

In the specific context of OPF analysis for monopolar DC networks, the SOC representation is adopted to address the nonlinear and non-convex nature of the problem. This choice is motivated by its capability to capture the hyperbolic relationship between voltage and power through a convex reformulation, which is crucial for ensuring the problem remains solvable using efficient conic optimization solvers [3]. To achieve this, an auxiliary variable representing the demanded current is introduced, enabling the transformation of the non-convex equations into a conic framework. This reformulation not only simplifies the mathematical complexity but also improves numerical stability and convergence properties during the optimization process [4].

Additionally, the SOC approach provides flexibility in integrating other system constraints, such as power flow limits and network operational bounds, into the optimization model. This makes it an ideal tool for solving large-scale network optimization problems, where both computational efficiency and solution accuracy are paramount [5]. The application of SOCs in OPF analysis thus represents a significant advancement in leveraging convex optimization techniques to address real-world challenges in power systems [6].

The general form of a second-order cone is mathematically described by

$$\|Ax + b\|_2 \leq C^{\top} x + d, \tag{5.1}$$

where $A \in \mathbb{R}^{m \times n}$ is a rectangular matrix, $b \in \mathbb{R}^m$ is a vector, $C \in \mathbb{R}^n$ is a column vector, $d \in \mathbb{R}$ is a scalar constant, and $x \in \mathbb{R}^n$ represents the decision variable vector. This structure ensures convexity, which is essential for guaranteeing global optimality and computational tractability in optimization problems.

A specific case of the second-order cone, known as the Lorentz cone or three-dimensional quadratic cone, is frequently used in applications [7]. This cone is defined by

$$\left\{(x, y, z) \in \mathbb{R}^3 \mid \sqrt{x^2 + y^2} \leq z\right\}, \tag{5.2}$$

where $(x, y, z)$ are the decision variables. The Lorentz cone serves as a geometric representation of quadratic constraints and is widely employed in convex optimization to model nonlinear relationships in a tractable manner.

Figure 5.1 illustrates the general structure of the Lorentz cone in the three-dimensional space, where in (5.2) the $z$ variable was set as 10.

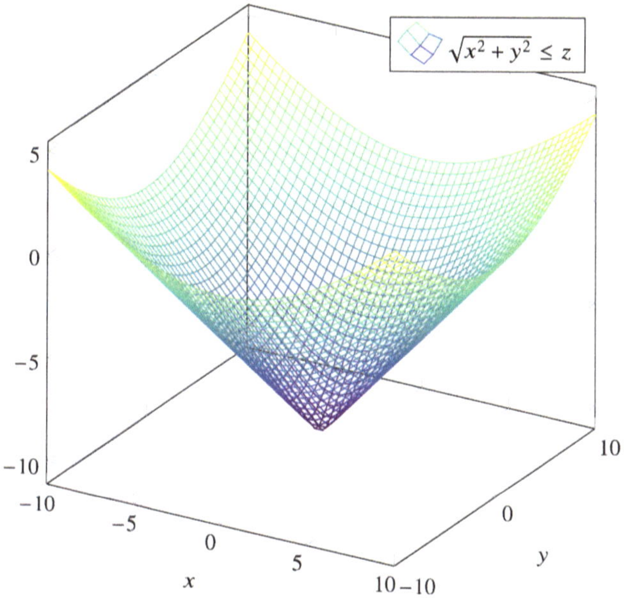


**Fig. 5.1** Lorentz cone in the three-dimensional space with $z = 10$

## 5.2 General Conic Approximation for a Hyperbolic Function

To obtain a conic representation of a hyperbolic function, consider the hyperbolic function defined in (5.3).

$$y = \frac{a}{x}, \tag{5.3}$$

where $x$ and $y$ are positive definite variables, and $a$ is a constant positive parameter. Note that the hyperbolic relation in (5.3) can be represented as follows:

$$\begin{aligned} a &= xy, \\ a &= \frac{1}{4}(x+y)^2 - \frac{1}{4}(x-y)^2, \\ 4a &= (x+y)^2 - (x-y)^2, \\ \left(2\sqrt{a}\right)^2 + (x-y)^2 &= (x+y)^2, \end{aligned} \tag{5.4}$$

which can be rewritten using the $l_2$-norm, i.e.,

$$\left\| \begin{matrix} 2\sqrt{a} \\ x-y \end{matrix} \right\| = x + y. \tag{5.5}$$

The equality constraint in (5.5) represents a conic constraint. However, it is inherently non-convex due to the strict imposition of equality, which prevents the use of efficient convex optimization techniques to solve the resulting problem. This non-convexity poses a significant challenge in optimization, as it typically leads to local minima and computational difficulties [1]. To address this limitation, a conic approximation of the hyperbolic relation in (5.3) is employed. Following the relaxation approach recommended in [8], strict equality is replaced with a convex inequality, enabling the transformation of the problem into a form suitable for second-order cone programming (SOCP).

The relaxed representation is given as follows:

$$\left\| \begin{matrix} 2\sqrt{a} \\ x - y \end{matrix} \right\| \leq x + y, \tag{5.6}$$

where the left-hand side represents the norm of a vector comprising the term $2\sqrt{a}$ and the difference $(x - y)$, and the right-hand side defines a linear bound involving the sum of $x$ and $y$. This relaxation effectively replaces the equality constraint with a second-order conic constraint, which is convex by definition.

## 5.3 Application to the OPF Problem

To obtain a recursive convex model via using the first conic approximation, we include a new group of variables that are associated with the constant power loads, i.e.,

$$i_{k,h}^{d} = \frac{p_{k,h}^{d}}{v_{k,h}}, \quad \{\forall k \in \mathbf{N}, \ \forall h \in \mathbf{H}\}, \tag{5.7}$$

which, by comparison to (5.5), can be relaxed as presented below.

$$\left\| \begin{matrix} 2\sqrt{p_{k,h}^{d}} \\ v_{k,h} - i_{k,h}^{d} \end{matrix} \right\| \leq v_{k,h} - i_{k,h}^{d}, \quad \{\forall k \in \mathbf{N}, \ \forall h \in \mathbf{H}\}. \tag{5.8}$$

Observe that in (5.8) it is assumed that $a = p_{k,h}^{d}$. For this reason, this parameter must be positive, i.e., a constant power load, since in the case of renewable generation, the hyperbolic relation between powers and voltages is relaxed as defined in Chap. 4.

Model (5.9) presents the first conic approximation to deal with the OPF problem in monopolar DC networks.

**First Conic Multiperiod OPF Approximation**

$$\text{O.F.:} \quad \min\{E_{\text{loss}}\} = \sum_{k\in\mathbf{N}}\sum_{m\in\mathbf{N}}\sum_{h\in\mathbf{H}} v_{k,h}^{t+1} G_{k,m} v_{m,h}^{t+1} \Delta_h \tag{5.9}$$

st:

$$\frac{p_{k,h}^{s}}{v_{k,h}^{t}} - \frac{p_{k,h}^{dg}}{v_{k,h}^{t}} - i_{k,h}^{d} = \sum_{m\in\mathbf{N}} G_{km} v_{m,h}^{t+1}, \qquad \{\forall k \in \mathbf{N},\ \forall h \in \mathbf{H}\}$$

$$\left\| \begin{matrix} 2\sqrt{p_{k,h}^{d}} \\ v_{k,h}^{t+1} + i_{k,h}^{d} \end{matrix} \right\| \le v_{k,h}^{t+1} - i_{k,h}^{d}, \qquad \{\forall k \in \mathbf{N},\ \forall h \in \mathbf{H}\}$$

$$r_{km} i_{km,h} = v_{k,h}^{t+1} - v_{m,h}^{t+1}, \qquad \{\forall km \in \mathbf{L},\ \forall h \in \mathbf{H}\}$$

$$\left| i_{km,h} \right| \le i_{km}^{\max}, \qquad \{\forall km \in \mathbf{L},\ \forall h \in \mathbf{H}\}$$

$$p_{k,h}^{s,\min} \le p_{k,h}^{s} \le p_{k,h}^{s,\max}, \qquad \{\forall k \in \mathbf{N},\ \forall h \in \mathbf{H}\}$$

$$p_{k,h}^{dg,\min} \le p_{k,h}^{dg} \le p_{k,h}^{dg,\max}, \qquad \{\forall k \in \mathbf{N},\ \forall h \in \mathbf{H}\}$$

$$v_{k}^{\min} \le v_{k,h}^{t+1} \le v_{k}^{\max}, \qquad \{\forall k \in \mathbf{N},\ \forall h \in \mathbf{H}\}$$

$$v_{k,h}^{t+1} = v_{k,h}^{t} = v_{\text{nom}}, \qquad \{\forall k \in \mathbf{S},\ \forall h \in \mathbf{H}\}.$$

## 5.4 Numerical Validation

The validation of the first conic OPF formulation in (5.9) is made by comparing its results with the exact OPF model solved in Chap. 3. The numerical implementations are made with the help of the `JuMP` optimization package and the `Ipopt` solver in the Julia programming language [9].

### *5.4.1 Test Feeder: 21-Bus System*

For this simulation scenario, we consider a radial distribution network consisting of 21 buses and 20 lines configured in a radial topology. The network includes a voltage-controlled node located at bus 1, which operates at a nominal voltage of

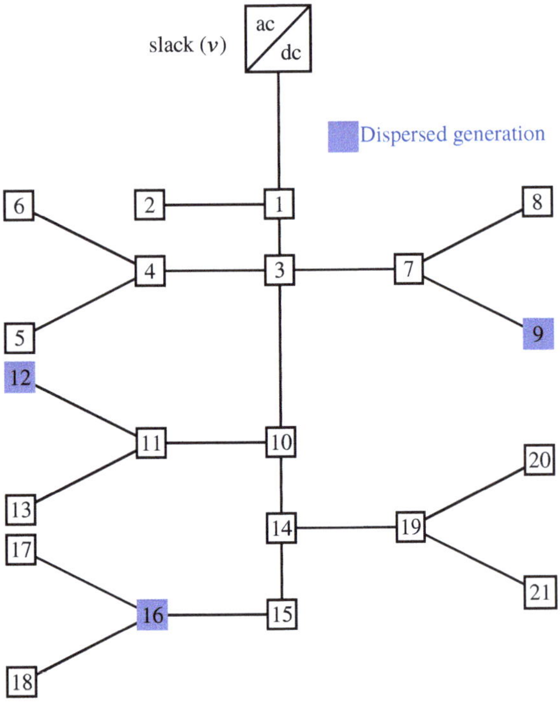


**Fig. 5.2** Electrical configuration for the 21-node test system

1 kV [10]. Additionally, three dispersed generators are integrated into the system at selected locations to analyze their impact on the network's performance. This configuration provides a representative test case for evaluating the behavior of distribution systems under varying load and generation conditions, particularly in scenarios involving distributed energy resources.

The electrical configuration of the 21-bus radial distribution network is depicted in Fig. 5.2, illustrating the connectivity and layout of the buses and lines. The radial topology is chosen for its prevalence in real-world distribution networks. Each distribution line in this test feeder is designed to support a maximum current of 600 A, ensuring sufficient capacity to handle typical operational conditions while allowing for transient surges within safety margins.

For this test system, the dispersed generation sources are located at nodes 9, 12, and 16 with sizes of 250, 300, and 200 kW, respectively. The electrical information regarding branch resistances and load consumptions is listed in Table 5.1.

### *5.4.2 Julia Software Implementation*

The first step in the Julia implementation consists of reaching the conductance nodal admittance matrix as presented below for the system under study:

**Table 5.1** Parametric information regarding branch resistances and peak load consumptions

| *Line parameters* | | | | | |
|---|---|---|---|---|---|
| (Line #) From | To | $R_{km}$ (Ω) | (Line #) From | To | $R_{km}$(Ω) |
| (Line 1) 1 | 2 | 0.0530 | (Line 11) 11 | 12 | 0.0790 |
| (Line 2) 1 | 3 | 0.0540 | (Line 12) 11 | 13 | 0.0780 |
| (Line 3) 3 | 4 | 0.0540 | (Line 13) 10 | 14 | 0.0830 |
| (Line 4) 4 | 5 | 0.0630 | (Line 14) 14 | 15 | 0.0650 |
| (Line 5) 4 | 6 | 0.0510 | (Line 15) 15 | 16 | 0.0640 |
| (Line 6) 3 | 7 | 0.0370 | (Line 16) 16 | 17 | 0.0740 |
| (Line 7) 7 | 8 | 0.0790 | (Line 17) 16 | 18 | 0.0810 |
| (Line 8) 7 | 9 | 0.0720 | (Line 18) 14 | 19 | 0.0780 |
| (Line 9) 3 | 10 | 0.0530 | (Line 19) 19 | 20 | 0.0840 |
| (Line 10) 10 | 11 | 0.0380 | (Line 20) 19 | 21 | 0.0820 |
| *Load consumptions* | | | | | |
| Node | $P$ (kW) | Node | $P$ (kW) | Node | $P$ (kW) |
| 1 | 0 | 2 | 70 | 3 | 0 |
| 4 | 36 | 5 | 4 | 6 | 36 |
| 7 | 0 | 8 | 32 | 9 | 80 |
| 10 | 0 | 11 | 45 | 12 | 68 |
| 13 | 10 | 14 | 0 | 15 | 22 |
| 16 | 23 | 17 | 43 | 18 | 34 |
| 19 | 9 | 20 | 21 | 21 | 21 |

```
using LinearAlgebra
using DataFrames
function Conductance(Lines,L,N)
    Gbus = zeros(N,N);
    for l = 1:L
        k = Lines.k[l,1];  m = Lines.m[l,1];
        rkm = Lines.Rkm[l,1];
        Gbus[k,m] = -1/rkm; Gbus[m,k] = -1/rkm;
        Gbus[k,k] = Gbus[k,k] + 1/rkm;
        Gbus[m,m] = Gbus[m,m] + 1/rkm;
    end
    return Gbus
end
Vb = 1; Pb = 0.1; Rb = Vb^2/Pb;
Ib = Pb/Vb; vmin = 0.90; vmax = 1.10;
Lines =  DataFrames.DataFrame([
    (1, 2,  0.0530, 0.60), (11, 12, 0.0790, 0.60),
    (1, 3,  0.0540, 0.60), (11, 13, 0.0780, 0.60),
    (3, 4,  0.0540, 0.60), (10, 14, 0.0830, 0.60),
    (4, 5,  0.0630, 0.60), (14, 15, 0.0650, 0.60),
    (4, 6,  0.0510, 0.60), (15, 16, 0.0640, 0.60),
    (3, 7,  0.0370, 0.60), (16, 17, 0.0740, 0.60),
    (7, 8,  0.0790, 0.60), (16, 18, 0.0810, 0.60),
    (7, 9,  0.0720, 0.60), (14, 19, 0.0780, 0.60),
```

```
    (3, 10, 0.0530, 0.60), (19, 20, 0.0840, 0.60),
    (10,11, 0.0380, 0.60), (19, 21, 0.0820, 0.60),]);
DataFrames.rename!(Lines, [:k, :m, :Rkm, :Ikm])
Nodes = DataFrames.DataFrame([
    (1,  1, 0,  0, 1000),(2,  1, 70, 0, 0),(3,  1, 0,  0, 0),
    (4,  1, 36, 0, 0),   (5,  1, 4,  0, 0),(6,  1, 36, 0, 0),
    (7,  1, 0,  0, 0),   (8,  1, 32, 0, 0),(9,  1, 80, 0, 250),
    (10, 1, 0,  0, 0),   (11, 1, 45, 0, 0),(12, 1, 68, 0, 300),
    (13, 1, 10, 0, 0),   (14, 1, 0,  0, 0),(15, 1, 22, 0, 0),
    (16, 1, 23, 0, 200), (17, 1, 43, 0, 0),(18, 1, 34, 0, 0),
    (19, 1, 9,  0, 0),   (20, 1, 21, 0, 0),(21, 1, 21, 0, 0),]);
DataFrames.rename!(Nodes, [:k, :Vk0, :Pk, :Pgmin, :Pgmax])
Lines.Rkm= Lines.Rkm/Rb;  Lines.Ikm = Lines.Ikm/Ib;
Nodes.Pk = Nodes.Pk/(Pb*1000); Nodes.Pgmin
      = Nodes.Pgmin/(Pb*1000);
Nodes.Pgmax = Nodes.Pgmax/Pb;
N = size(Nodes,1); L = size(Lines,1);
Gbus = Conductance(Lines,L,N);
```

Once the conductance matrix is constructed with the test system information, the optimization model that represents the first conic approximation in (5.9) is implemented using a function called `FCA(Gbus,Nodes,Lines,vt,N,L,vmin,vmax)`. Note that this function was built to implement the recursive solution of the optimization model using the iterative updating of the voltage variables until reaching the desired convergence.

```
using JuMP, Ipopt
function FCA(Gbus,Nodes,Lines,vt,N,L,vmin,vmax)
    elements = 1:N;  branches = 1:L; a = 0;
    Pd = Nodes.Pk; Pgmin = Nodes.Pgmin;
    Pgmax = Nodes.Pgmax; Ilmax = Lines.Ikm;
    ROPF = Model(Ipopt.Optimizer)
    @variable(ROPF,pg[i in elements])
    @variable(ROPF,v[i in elements])
    @variable(ROPF,il[l in branches])
    @variable(ROPF,id[l in elements])
    @objective(ROPF,Min, sum(v[i]*sum(Gbus[i,j]*v[j]
    for j in elements) for i in elements))
    for i = 1:N
        @constraint(ROPF,pg[i] >= Pgmin[i])
        @constraint(ROPF,pg[i] <= Pgmax[i])
        @constraint(ROPF,v[i] <= vmax)
        @constraint(ROPF,v[i] >= vmin)
        @NLconstraint(ROPF, pg[i]/vt[i] - id[i]
        == sum(Gbus[i,j]*v[j] for j in elements))
        if Pd[i] == 0
            @constraint(ROPF,id[i] == 0)
        else
            @NLconstraint(ROPF,sqrt((2*sqrt(Pd[i]))^2 +
            (v[i]-id[i])^2) <= v[i]+id[i])
        end
    end
    @constraint(ROPF,v[1] == 1);
```

```
    @constraint(ROPF,sum(pg[i] for i = 2:N) == a*sum(Pd));
    for l = 1:L
        k = Lines.k[l,1]; m = Lines.m[l,1]; rkm = Lines.Rkm[l,1];
        @constraint(ROPF,il[l] == (v[k] - v[m])/rkm)
        @NLconstraint(ROPF,-Ilmax[l] <= il[l] <= Ilmax[l])
    end
    JuMP.optimize!(ROPF)
    pg = [value(pg[i]) for i in elements];
    vol = [value(v[i]) for i in elements];
    il = [value(il[j]) for j in branches];
    fobjval = @show objective_value(ROPF)*Pb;
    return vol, il, pg, fobjval;
end
```

Note that in this implementation, a new constraint was added in order to validate different penetration of dispersed generation levels, i.e., `@constraint(ROPF, sum(pg[i] for i = 2:N) == a*sum(Pd));`, which means that a is a constant that defines which is the maximum of power injection permitted for all the dispersed generation sources. Finally, the next script is used to implement the recursive solution of the OPF problem using the first conic approximation.

```
epsilon = 1e-10; itermax = 100;
function RCS(Gbus,Nodes,Lines,epsilon,itermax)
    vt = ones(N,1);
    for t = 1:itermax
        global vol,il, pg, objval = FCA(Gbus,Nodes,Lines,
                                        vt,N,L,vmin,vmax)
        if maximum(abs.(abs.(vol) - abs.(vt)))< epsilon
            print("Number of iterations = ",t)
            break
        else
            vt = vol
        end
    end
    return vol,il, pg, objval;
end
vol,il,pg,fobjval = RCS(Gbus,Nodes,Lines,epsilon,itermax)
Report1 = DataFrames.DataFrame(;
Bus = 1:N, Voltage = vol*Vb, Pgen =pg*Pb);
Report2 = DataFrames.DataFrame(;
Lines = 1:L, From = Lines.k, To = Lines.m,
Current = il*Ib, Loss = Pb*Lines.Rkm.*(il).^2,);
```

Observe that `RCS(Gbus,Nodes,Lines,epsilon,itermax)` implements the recursive solution until the desired convergence is reached.

Some of the numerical results reached in the 21-bus grid for different values of the a parameter are listed in Table 5.2. It is worth mentioning that the proposed first conic approach takes about three iterations when no dispersed generation is considered. When distributed generation is dispatched, the 20%, 40%, and 60% penetration levels require seven iterations to solve the OPF problem.

**Table 5.2** Numerical results in the 21-bus grid with the first conic approximation

| a-value (%) | Power gen. (kW) | Losses (kW) | Min. voltage (p.u.) |
|---|---|---|---|
| 0 | $\begin{bmatrix} p_1^s \\ p_9^{dg} \\ p_{12}^{dg} \\ p_{16}^{dg} \end{bmatrix} = \begin{bmatrix} 581.6304 \\ 0.0000 \\ 0.0000 \\ 0.0000 \end{bmatrix}$ | 27.6034 | $v_{17} = 0.9211$ |
| 20 | $\begin{bmatrix} p_1^s \\ p_9^{dg} \\ p_{12}^{dg} \\ p_{16}^{dg} \end{bmatrix} = \begin{bmatrix} 456.8540 \\ 0.0000 \\ 11.8622 \\ 98.9378 \end{bmatrix}$ | 13.6539 | $v_{20} = 0.9565$ |
| 40 | $\begin{bmatrix} p_1^s \\ p_9^{dg} \\ p_{12}^{dg} \\ p_{16}^{dg} \end{bmatrix} = \begin{bmatrix} 339.0133 \\ 22.4621 \\ 69.7882 \\ 129.3495 \end{bmatrix}$ | 6.6132 | $v_{20} = 0.9705$ |
| 60 | $\begin{bmatrix} p_1^s \\ p_9^{dg} \\ p_{12}^{dg} \\ p_{16}^{dg} \end{bmatrix} = \begin{bmatrix} 224.6615 \\ 83.5013 \\ 102.5788 \\ 146.3199 \end{bmatrix}$ | 3.0614 | $v_{20} = 0.9809$ |

Numerical results in Table 5.2 shows that:

i. The minimum voltage profile is presented at bus 17 when no dispersed generation is considered, which means a voltage regulation value of 7.8899%; however, for any level of distributed generation penetration, the worst voltage appears at bus 20. Note that for a = 60%, the voltage regulation of the 21-bus grid improves to 1.91%.
ii. Regarding power losses, it is observed that the worst case occurs when no dispersed generation is dispatched with a value of 27.6034%; however, for a = 20%, this value is reduced to about 50.5354%, i.e., 13.9495 kW. In the case of a = 60%, the power loss improvement is about 88.9093%.
iii. Regarding the power dispatch of the distributed generation sources, the most sensitive node for power injection corresponds to bus 16, while the less sensitive generation node is at bus 9. It can be explained by the fact that node 16 is far from the substation in the area where the worst voltage profiles appear due to the radial grid configuration. In contrast, bus 9 is near the substation bus, and its influence on the total grid power losses is lower.

## 5.5 Conclusion

The integration of dispersed generation in the 21-bus radial distribution network demonstrates significant improvements in both efficiency and stability. With increasing penetration levels of distributed generation, the voltage regulation across the network is enhanced, reducing the worst-case voltage drop from 7.89% (without generation) to 1.91% (at 60% penetration). Additionally, power losses are drastically minimized, with an 88.91% reduction achieved at maximum distributed generation levels. The results also underscore the importance of strategic generator placement, as bus 16 exhibits the highest impact due to its location in areas with poor voltage profiles, while bus 9 shows minimal impact due to its proximity to the substation. These findings highlight the critical role of dispersed generation in optimizing the performance of radial distribution systems.

## References


1. Boyd, S., & Vandenberghe, L. (2004). *Convex optimization*. Cambridge University Press.
2. Ben-Tal, A., & Nemirovski, A. (2001). *Lectures on Modern Convex Optimization: Analysis, Algorithms, and Engineering Applications*. Society for Industrial and Applied Mathematics.
3. Nazir, F. U., Pal, B., & Jabr, R. (2021). Approximate load models for conic OPF solvers. In *2021 IEEE Power & Energy Society General Meeting (PESGM)* (pp. 1–1). IEEE.
4. Gan, L., & Low, S. H. (2014). Optimal power flow in direct current networks. *IEEE Transactions on Power Systems, 29*(6), 2892–2904.
5. Vandenberghe, L., & Boyd, S. (1996). Semidefinite programming. *SIAM Review, 38*(1), 49–95.
6. Chen, Y., Qi, H., Li, H., Xu, H., Yang, Q., & Chen, Q. (2022). Convex-optimization-based power-flow calculation method for offshore wind systems. *Energies, 15*(20), 7717.
7. Huang, C.-H., & Chen, J.-S. (2019). On unitary elements defined on Lorentz cone and their applications. *Linear Algebra and its Applications, 565*, 1–24.
8. Farivar, M., & Low, S. H. (2013). Branch flow model: Relaxations and convexification—part I. *IEEE Transactions on Power Systems, 28*(3), 2554–2564.
9. Lubin, M., Dowson, O., Dias Garcia, J., Huchette, J., Legat, B., & Vielma, J. P. (2023). JuMP 1.0: Recent improvements to a modeling language for mathematical optimization. *Mathematical Programming Computation, 15*(3), 581–589.
10. Montoya, O. D., Gil-González, W., & Hernández, J. C. (2023). Optimal power flow solution for bipolar DC networks using a recursive quadratic approximation. *Energies, 16*(2), 589.

# Chapter 6
# Semidefinite Programming Model

**Abstract** This chapter explores the application of semidefinite programming (SDP) to solve the optimal power flow (OPF) problem in monopolar multiterminal high-voltage direct current (MT-HVDC) systems. By leveraging the convex relaxation capabilities of SDP, the inherently non-convex OPF problem is transformed into a convex framework, ensuring computational efficiency and global optimality. The chapter presents the theoretical foundations of SDP, details the reformulation of the OPF problem using matrix-based decision variables, and demonstrates the use of SDP to optimize power generation, minimize power losses, and maintain voltage stability. Numerical validations on an 11-node MT-HVDC test system highlight the method's effectiveness in maintaining voltages within acceptable limits, achieving balanced power generation, and minimizing system losses. The results also confirm the efficient distribution of line currents, ensuring operational safety and thermal compliance. This study underscores the versatility and robustness of SDP as a critical tool for addressing complex, large-scale optimization problems in modern power systems, particularly in networks with renewable energy integration and dynamic operational conditions.

**Keywords** Semidefinite programming · Convex relaxation · Optimal power flow · MT-HVDC systems · Voltage stability

## 6.1 Semidefinite Programming Theory

Semidefinite programming (SDP) is a powerful branch of convex optimization that has garnered significant attention in recent years due to its versatility and ability to solve a wide range of optimization problems [1]. Unlike traditional convex optimization methods, SDP has the unique capability to find globally optimal solutions even for certain non-convex problems, provided that specific conditions, such as convex relaxations, are satisfied [2]. This has made SDP a critical tool in various domains, including optimization, control theory, signal processing, communication systems, and power system analysis [3].

© The Author(s), under exclusive license to Springer Nature Switzerland AG 2026

O. D. Montoya Giraldo et al., *Optimal Power Flow Methods for DC Networks*, Power Systems, https://doi.org/10.1007/978-3-032-05734-1_6

The general formulation of an optimization problem using SDP can be expressed as [4]

$$
\begin{aligned}
&\min z = \mathrm{tr}(CX) \\
&\text{subject to:} \\
&\qquad A_j X = B_j, \quad \forall\, j \in \{1, 2, \ldots, m\}, \\
&\qquad X \succeq 0,
\end{aligned}
\tag{6.1}
$$

where $z$ represents the objective function, $C \in \mathbb{R}^{n\times n}$ is a square cost matrix, and $X \in \mathbb{R}^{n\times n}$ is the decision variable matrix. The trace operator, denoted as $\mathrm{tr}(\cdot)$, sums the diagonal elements of its matrix argument. Additionally, $A_j \in \mathbb{R}^{m\times n}$ and $B_j \in \mathbb{R}^m$ are the parameter matrices and vectors, respectively. The constraint $X \succeq 0$ ensures that the matrix $X$ is positive semidefinite, a critical requirement in SDP formulations.

SDP is particularly advantageous because it allows for the convex relaxation of problems that are otherwise non-convex [1]. This means that SDP can approximate or reformulate non-convex problems into a convex form, enabling the use of efficient numerical solvers to compute globally optimal solutions. Applications of SDP span a wide range of fields. For instance:

- **Control Systems:** SDP is extensively used in control system design, including stability analysis, robust control, and model predictive control [5]. It provides tools to derive Lyapunov functions and solve linear matrix inequality (LMI) problems, which are fundamental for ensuring system stability under uncertain conditions. Applications include the design of feedback controllers, disturbance rejection schemes, and fault-tolerant systems. SDP's ability to handle multivariable systems and constraints makes it a powerful framework for modern control theory.
- **Power Systems:** In power systems, SDP is used for optimal power flow (OPF) problems, voltage stability analysis, and state estimation, where its ability to handle large-scale, non-convex formulations makes it invaluable [6]. It enables the incorporation of renewable energy sources, distributed generation, and energy storage systems into the optimization process. SDP has also been employed to address uncertainties in load demand and generation, ensuring reliable and efficient operation of complex networks [4]. Additionally, its convex relaxation techniques provide tight approximations for highly nonlinear power system models.
- **Signal Processing and Communication:** SDP finds applications in beamforming, resource allocation, and filter design, enabling optimal solutions in high-dimensional settings [7, 8]. For instance, it is used to design robust beamforming algorithms that minimize interference and maximize signal quality in wireless communication networks. Resource allocation problems, such as bandwidth and power distribution, benefit from SDP's ability to enforce fairness and efficiency simultaneously. In filter design, SDP ensures optimal trade-offs

between noise suppression and signal fidelity, making it a cornerstone in modern signal processing.

- **Economics and Finance:** SDP is applied in portfolio optimization and resource allocation problems, ensuring risk minimization while adhering to constraints [9]. It allows for the modeling of complex financial objectives, such as maximizing returns while accounting for market uncertainties and transaction costs. In resource allocation, SDP ensures equitable distribution of resources under multi-objective and multi-constraint scenarios. Its ability to model quadratic cost functions and covariance matrices makes it highly effective for financial risk analysis and decision-making [10].

The matrix-based nature of SDP and its reliance on positive semidefiniteness provide it with flexibility and robustness when handling constraints that involve quadratic or higher-order terms. Furthermore, state-of-the-art solvers such as `CVX`, `MOSEK`, and `SDPT3` have been developed to handle large-scale SDP problems efficiently, further extending its practical utility [11].

In summary, semidefinite programming stands as a cornerstone of modern optimization methodologies, offering a robust and efficient framework for solving complex problems in diverse engineering and scientific disciplines [2]. Its theoretical elegance and practical applicability continue to expand its influence across emerging fields [1].

Figure 6.1 illustrates an example of positive semidefinite matrix $X \in \mathbb{R}^{2\times 2}$, which can be defined as

$$X = \begin{bmatrix} x_{11} & x_{12} \\ x_{12} & x_{22} \end{bmatrix} \succeq 0. \tag{6.2}$$

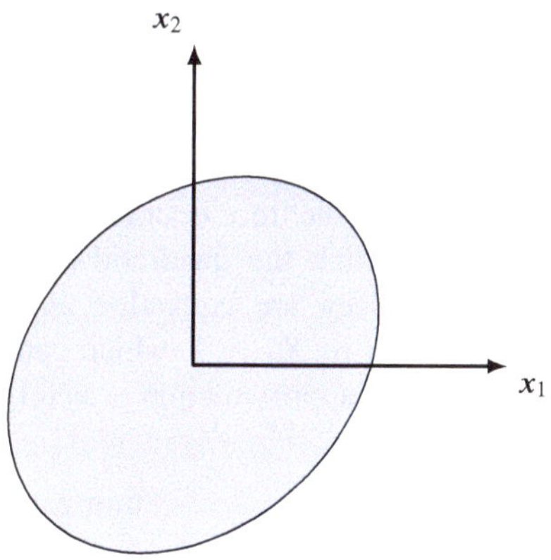


**Fig. 6.1** Example of positive semidefinite matrix $X \in \mathbb{R}^{2\times 2}$

## 6.2 General Semidefinite Approximation

Let us define a typical quadratic problem as follows [1]:

$$\begin{aligned} &\min z = \frac{1}{2} x^\top C x \\ &\text{Subject to:} \\ &x^\top A_j x = B_j \ \forall \ j \in \{1, 2, 3, ..., n\}, \end{aligned} \tag{6.3}$$

where $x \in \mathbb{R}^{1 \times n}$ is the variable decision vector and $C \succeq 0 \in \mathbb{R}^{1 \times n}$ is a semidefinite matrix.

Note that the optimization model (6.3) is a non-convex model because of its quadratic restriction of equality. However, it is possible to transform it into a convex model by applying a semidefinite approximation [3]. Initially, define the matrix decision variable as follows:

$$X = x x^\top. \tag{6.4}$$

Note that the matrix decision variable $X = x x^\top \in \mathbb{R}^{n \times n}$ is obtained making the product of the variable decision vector $x \in \mathbb{R}^{n \times 1}$ by its transpose.

Now the quadratic problem (6.3) takes the following form:

$$\begin{aligned} &\min z = \frac{1}{2} \mathrm{tr}(C X) \\ &\text{Subject to:} \\ &A_j X = B_j \ \forall \ j \in \{1, 2, 3, ..., n\} \\ &\mathrm{rn}(X) = 1 \\ &X \succeq 0, \end{aligned} \tag{6.5}$$

where $\mathrm{tr}(\cdot)$ is the trace operator of argument.

Observe that the quadratic problem (6.3) has been transformed into an SDP problem. They are equivalent and not convex. Nevertheless, the complicating constraint is $\mathrm{rn}(X) = 1$, which can be relaxed of the problem. Hence, the general semidefinite approximation is as follows:

$$\begin{aligned} &\min z = \frac{1}{2} \mathrm{tr}(C X) \\ &\text{Subject to:} \\ &A_j X = B_j \ \forall \ j \in \{1, 2, 3, ..., n\} \\ &X \succeq 0. \end{aligned} \tag{6.6}$$

On the other hand, it is necessary to recover the original variables; thus the matrix decision variable $X$ can be written as follows:

$$X = \sum_{j}^{n} \lambda_j \phi_j \phi_j^\top, \tag{6.7}$$

where $\lambda_j$ denotes the eigenvalue $j$ of the matrix $X$ and $\phi_j$ is its respective eigenvector. Since $\mathrm{rn}(X) \approx 1$, there is an eigenvalue much greater than the rest of them. In other words, the maximum eigenvalue $\lambda_k \gg \lambda_j \ \forall \ j \neq k$, where $\lambda_k = \max\{\lambda_1, ..., \lambda_n\}$. Hence, the original decision variables are recovered:

$$x = \sqrt{(}\lambda_k)\phi_j, \tag{6.8}$$

where the $l_2$-norm of $\phi_j$ is equal to 1, i.e., $\left\|\phi_j\right\| = 1$.

## 6.3 Application to the OPF Problem

To obtain a linear matrix model by using the semidefinite approximation, we define the voltages of the OPF problem, as follows:

$$X_h = v_h v_h^\top, \ \{\forall h \in \mathbf{H}\}, \tag{6.9}$$

where $v_h \in \mathbf{N}$ is the voltage vector, i.e., $v_h = [v_{1,h}, ..., v_{n,h}]^\top$.

The quadratic objective function $E_{\mathrm{loss}}$ presented in (2.26) can be transformed into a linear objective function as follows:

$$E_{\mathrm{loss}} = \sum_{h \in \mathbf{H}} \mathrm{tr}\left(G_{\mathrm{bus}} X_h\right) \Delta_h. \tag{6.10}$$

Similarly, the sum of power flow at node $k$ can be rewritten as a linear equation using the matrix decision variable, as follows:

$$v_{k,h} \sum_{m \in \mathbf{N}} g_{km} v_{m,h} = \mathrm{diag}_k\left(G_{\mathrm{bus}} X(h)\right), \ \{\forall k \in \mathbf{N}, \ \forall h \in \mathbf{H}\}, \tag{6.11}$$

where $\mathrm{diag}_k(\cdot)$ is the operator that takes the position $k$ of the diagonal of the matrix.

The power flow by line $km$ and its maximum capacity can be presented as follows:

$$p_{km,h} = g_{km}\left(v(h)_k^2 - v(h)_k v(h)_m\right), \ \{\forall km \in \mathbf{L}, \ \forall h \in \mathbf{H}\} \tag{6.12}$$

$$p_{km,h} = g_{km}\left(X_{kk,h}^2 - X_{km,h}\right), \ \{\forall km \in \mathbf{L}, \ \forall h \in \mathbf{H}\} \tag{6.13}$$

$$\left|p_{km,h}\right| \leq p_{km}^{\max}, \ \{\forall km \in \mathbf{L}, \ \forall h \in \mathbf{H}\}. \tag{6.14}$$

Model (6.15) models the SDP approximation to deal with the OPF problem in monopolar DC networks.

**SDP Multiperiod OPF Approximation**

$$\text{O.F.:} \quad \min\{E_{\text{loss}}\} = \sum_{h \in \mathbf{H}} \operatorname{tr}(G_{\text{bus}} X_h)\, \Delta_h. \tag{6.15}$$

st:

$$
\begin{aligned}
& P_h^s + P_h^{dg} - P_h^d = \operatorname{diag}(G_{\text{bus}} X(h)), && \{\forall h \in \mathbf{H}\} \\
& p(h)_{km} = g_{km}\left(X_{kk,k} - X_{km,h}\right), && \{\forall km \in \mathbf{L}, \ \forall h \in \mathbf{H}\} \\
& \left|P_{km,h}\right| \leq P_{km}^{\max}, && \{\forall km \in \mathbf{L}, \ \forall h \in \mathbf{H}\} \\
& P_h^{s,\min} \leq P_h^s \leq P_h^{s,\max}, && \{\forall h \in \mathbf{H}\} \\
& P_h h^{dg,\min} \leq P_h^{dg} \leq P_h^{dg,\max}, && \{\forall k \in \mathbf{N}, \ \forall h \in \mathbf{H}\} \\
& 1_N \left(v_k^{\min}\right)^2 \leq X(h) \leq 1_N \left(v_k^{\max}\right)^2, && \{\forall h \in \mathbf{H}\} \\
& X(h) = X(h)^\top \succeq 0, && \{\forall h \in \mathbf{H}\} \\
& X_{kk,h} = v_{\text{nom}}, && \{\forall k \in \mathbf{S}, \ \forall h \in \mathbf{H}\}.
\end{aligned}
$$

## 6.4 Numerical Validation

The validation of the SDP formulation in (6.15) is made by comparing its results with the exact OPF model solved in Chap. 3. The numerical implementations are made with the help of the `JuMP` optimization package and the `Mosek` solver in the Julia programming language [12]. The test system used in this part is the 11-bus MT-HVDC shown in Sect. 4.3.1

### *6.4.1 Implementation of SDP Formulation*

This implementation initializes with the same function `Conductance(Lines,L,N)`, which was defined in Sect. 4.3.2. Remember that this function allows reaching the value of the conductance matrix for the test system. Once the information of the test system was uploaded and the $G_{\text{bus}}$ matrix is computed, then the nonlinear solver

Mosek using the JuMP optimization environment is employed to solve the OPF model (6.15) as described below:

```
using MosekTools, JuMP, LinearAlgebra
function solveSDP(Gbus,Nodes,Lines,N,L,vmin,vmax)
    elements = 1:N;  branches = 1:L;
    Pd = Nodes.Pk; Pgmin = Nodes.Pgmin;
    Pgmax = Nodes.Pgmax; Ilmax = Lines.Ikm;

    SDP = Model(optimizer_with_attributes(Mosek.Optimizer))
    set_silent(SDP)
    @variable(SDP,Pg[elements])
    @variable(SDP,X[1:N,1:N],PSD)
    @variable(SDP,il[branches])
    @objective(SDP,Min, tr(Gbus * X))
   # W = Gbus * X

    @constraint(SDP,Pg-Pd == diag(Gbus * X))

    #for i in elements
       # @constraint(SDP,Pg[i]-Pd[i] == W[i,i])
    @constraint(SDP,Pg <= Pgmax)
    @constraint(SDP,Pg >= Pgmin)
    for i in elements
        for j in elements
            @constraint(SDP,X[i,j] <= vmax^2)
            @constraint(SDP,X[i,j] >= vmin^2)
        end
    end
    @constraint(SDP,X[3,3] == 1)
    for l in branches
        k = Lines.k[l,1];  m = Lines.m[l,1];
        @constraint(SDP,il[l] == 1/((Lines.Rkm[l,1])^2)*
        (X[k,k]-2*X[k,m]+X[m,m]))
        @constraint(SDP,il[l] <= Ilmax[l]^2)
    end

    optimize!(SDP)
    status = JuMP.termination_status(SDP)
    vol = JuMP.value.(X)
    vol = diag(vol)
    for i in elements
        vol[i] = sqrt(vol[i])
    end
    fobjval = JuMP.objective_value(SDP)
    pg = JuMP.value.(Pg)
    il = JuMP.value.(il)
    return vol, il, pg, fobjval, status;
end

vol, il, pg, fobjval,
status = solveSDP(Gbus,Nodes,Lines,N,L,vmin,vmax)
Report1 = DataFrames.DataFrame(
Bus = 1:N, Voltage = vol*Vb, Pgen =collect(pg)*Pb);
```

```
Report2 = DataFrames.DataFrame(;
Lines = 1:L, From = Lines.k, To = Lines.m,
Current = collect(il)*Ib, Loss = Pb*Lines.Rkm.*collect(il).^2,);
```

## 6.5 Analysis of Results

After executing the SDP approximation, the results shown in Table 6.1 are obtained.

The results obtained using the SDP approximation for the 11-node MT-HVDC system, as shown in Table 6.1, reveal the following key observations:

- The node voltages are maintained close to the nominal value of 380 kV, with minor deviations observed at nodes 6, 7, 8, 9, 10, and 11. The lowest voltage is recorded at node 8, with a value of 371.33 kV, representing a deviation of approximately 2.28%. These deviations are within acceptable voltage regulation limits, indicating that the SDP approximation effectively maintains system stability while adhering to operational constraints.
- The power generation values for the source nodes demonstrate a balanced contribution across the system, with node 5 contributing the highest power at 1632.22 MW. The total power losses for the system amount to 70.43 MW, which is a reasonable value considering the size of the system and the distribution of loads. This highlights the efficiency of the SDP approach in minimizing power losses while ensuring sufficient power supply to all nodes.
- The line currents indicate the flow of power throughout the network, with line 12 carrying the highest current at 1.7624 kA, followed by line 10 at 1.6936 kA. The distribution of currents aligns with the voltage and power flow characteristics, confirming that the system operates within its thermal limits. The

**Table 6.1** Voltage, power, and current outputs for the 11-node MT-HVDC system

| Node | $v_k$ (kV) | $p_k^s$ (MW) | $p_{loss}$ (MW) | Line | $i_{km}$ (kA) | Line | $i_{km}$ (kA) |
|---|---|---|---|---|---|---|---|
| 1 | 380.4174 | 281.1425 | | 1 | 0.0064 | 12 | 1.7624 |
| 2 | 380.3928 | 1140.8443 | | 2 | 0.0047 | 13 | 0.2896 |
| 3 | 380.0000 | 1035.6739 | | 3 | 0.7373 | 14 | 0.5851 |
| 4 | 380.4372 | 680.5444 | | 4 | 1.4985 | 15 | 1.3264 |
| 5 | 380.4319 | 1632.2248 | | 5 | 1.5071 | 16 | 0.7573 |
| 6 | 376.8414 | – | 70.4289 | 6 | 1.5272 | 17 | 0.4747 |
| 7 | 375.4949 | – | | 7 | 1.1983 | – | – |
| 8 | 371.3299 | – | | 8 | 0.8945 | – | – |
| 9 | 374.7754 | – | | 9 | 0.8897 | – | – |
| 10 | 376.9934 | – | | 10 | 1.6936 | – | – |
| 11 | 373.7699 | – | | 11 | 0.8345 | – | – |

results demonstrate that the SDP approximation effectively optimizes current flow, avoiding overloading and maintaining operational safety.

In summary, the SDP approximation provides a robust and efficient solution for the 11-node MT-HVDC system. It ensures voltage stability, minimizes power losses, and maintains line currents within thermal limits. Despite minor approximations, the objective function value remains consistent with alternative methods, reinforcing the reliability of the SDP approach for large-scale power system optimization.

## 6.6 Conclusions

The SDP approximation demonstrates its robustness and effectiveness in solving the OPF problem for the 11-node MT-HVDC system. The results validate that the SDP-based approach maintains voltage stability within acceptable regulation limits, with deviations of less than 2.5% from the nominal voltage values. Furthermore, the method ensures a balanced power generation profile while achieving minimal power losses of 70.43 MW, which accounts for a small fraction of the total generation. The analysis of line currents confirms that the network operates well within its thermal limits, avoiding potential overloading scenarios. Overall, the SDP approach provides a computationally efficient and globally optimal solution, demonstrating its suitability for large-scale and complex power system optimization tasks. Its flexibility and scalability make it a valuable tool for addressing challenges in modern power systems, particularly in networks with renewable energy integration and dynamic operational conditions.

## References


1. Boyd, S., & Vandenberghe, L. (2004). *Convex optimization*. Cambridge University Press.
2. Garcés, A. (2021). *Mathematical Programming for Power Systems Operation with Applications in Python: From Theory to Applications in Python*. Wiley.
3. Gil-González, W., Montoya, O. D., Holguín, E., Garces, A., & Grisales-Noreña, L. F. (2019). Economic dispatch of energy storage systems in DC microgrids employing a semidefinite programming model. *Journal of Energy Storage, 21*, 1–8.
4. Garcés Ruiz, A. (2020). *Convex optimization, applications in operation and dynamics of power systems (in Spanish)*. Universidad Tecnológica de Pereira.
5. Siddhartha, V., & Mahapatra, S. (2022). Design of a state derivative optimal control law using LMI technique for diving motion of autonomous underwater vehicle. In *2022 2nd International Conference on Artificial Intelligence and Signal Processing (AISP)* (pp. 1–5). IEEE.
6. Bai, X., Wei, H., Fujisawa, K., & Wang, Y. (2008). Semidefinite programming for optimal power flow problems. *International Journal of Electrical Power & Energy Systems, 30*(6–7), 383–392.
7. Tan, P. H., & Rasmussen, L. K. (2001). The application of semidefinite programming for detection in CDMA. *IEEE Journal on Selected Areas in Communications, 19*(8), 1442–1449.

8. Wu, S.-P., Boyd, S., & Vandenberghe, L. (1996). Fir filter design via semidefinite programming and spectral factorization. In *Proceedings of 35th IEEE Conference on Decision and Control (CDC-96)* (vol. 1, pp. 271–276). IEEE.
9. Costa, O. L.V., de Oliveira Ribeiro, C., Rego, E. E., Stern, J. M., Parente, V., & Kileber, S. (2017). Robust portfolio optimization for electricity planning: An application based on the Brazilian electricity mix. *Energy Economics, 64*, 158–169.
10. Gepp, A., Harris, G., & Vanstone, B. (2019). Financial applications of semidefinite programming: A review and call for interdisciplinary research. *Accounting & Finance, 60*(4), 3527–3555.
11. Grant, M., & Boyd, S. (2014). CVX: Matlab software for disciplined convex programming, version 2.1. https://cvxr.com/cvx.
12. Lubin, M., Dowson, O., Dias Garcia, J., Huchette, J., Legat, B., & Vielma, J. P. (2023). JuMP 1.0: Recent improvements to a modeling language for mathematical optimization. *Mathematical Programming Computation, 15*, 581–589.

# Chapter 7
# Second Conic Approximation

**Abstract** This chapter presents the application of second-order cone programming (SOCP) to solve the optimal power flow (OPF) problem in direct current (DC) distribution networks. The proposed approach reformulates nonlinear constraints, particularly those involving products of positive variables, into convex second-order conic constraints, ensuring computational efficiency and maintaining solution accuracy. The mathematical formulation begins with a generic nonlinear constraint and introduces auxiliary variables to approximate quadratic terms, leveraging conic relaxations to handle non-convexities effectively. The methodology is validated using the 33-bus DC test system, which includes distributed generation (DG) units at nodes 12, 15, and 31. Numerical results demonstrate the reliability and effectiveness of the proposed SOCP approach. Daily energy losses were reduced by 43.98% when DG units were operational, highlighting the significant impact of integrating renewable energy sources. The maximum line loadability and thermal constraints across all branches were also respected, ensuring safe and efficient network operation. Furthermore, detailed power generation analyses reveal that DG units operate below their maximum capacities due to thermal limitations and optimal coordination with the slack generator. This strategic operation minimizes energy losses while maintaining voltage and current constraints within permissible limits. The chapter concludes with a discussion on the computational advantages and practical implications of the SOCP-based OPF solution for future energy systems, underscoring its potential for scalability and real-world applications.

**Keywords** Second-order cone programming · Convex optimization · Optimal power flow · DC distribution networks · Distributed generation

## 7.1 Second-Order Cone Programming Approximation

Second-order cone programming (SOCP) is a powerful optimization tool for approximating nonlinear equations, particularly those involving products between positive variables [1]. To demonstrate the application of SOCP, let us consider the nonlinear equation described in (7.1), which represents a generic constraint with

© The Author(s), under exclusive license to Springer Nature Switzerland AG 2026 
O. D. Montoya Giraldo et al., *Optimal Power Flow Methods for DC Networks*,
Power Systems, https://doi.org/10.1007/978-3-032-05734-1_7

quadratic terms [2]:

$$y_k - a_k = \sum_{m \in \mathbf{N}} b_{km} x_k x_m, \ \forall k \in \mathbf{N}. \tag{7.1}$$

Here, $y_k$ is a linear variable, $a_k$ and $b_{km}$ are constant parameters, and $x_k$ and $x_m$ are continuous, positive variables. The challenge lies in handling the product $x_k x_m$, which introduces nonlinearity into the equation.

The key idea of the SOCP approximation is to reformulate Eq. (7.1) into an affine representation by introducing new auxiliary variables $z_{km}$ [3]. This reformulation is shown in Eq. (7.2):

$$y_k - a_k = \sum_{m \in \mathbf{N}} b_{km} z_{km}, \ \forall k \in \mathbf{N}. \tag{7.2}$$

In this expression, the quadratic product $x_k x_m$ is replaced by the auxiliary variable $z_{km}$, simplifying the structure of the original equation. However, to ensure that the approximation is accurate, specific conditions must be imposed on the variables $z_{km}$.

To achieve this, the relationship between $z_{km}$ and the original variables is defined through the following algebraic operations [4]:

$$\begin{aligned} z_{km} &= x_k x_m, \\ z_{kk} &= x_k x_k = x_k^2, \\ z_{mm} &= x_m x_m = x_m^2, \\ z_{km}^2 &= (x_k x_m)^2 = z_{kk} z_{mm}. \end{aligned} \tag{7.3}$$

Equation (7.3) ensures that the auxiliary variables $z_{km}$ accurately represent the products between the original variables. To incorporate these relationships into the optimization model, the second-order cone equivalent for the product of two positive variables is utilized [2]. This can be expressed using the $l_2$-norm, as follows:

$$\left\| \begin{matrix} 2z_{km} \\ z_{kk} - z_{mm} \end{matrix} \right\| = z_{kk} + z_{mm}. \tag{7.4}$$

The equality constraint in (7.4) represents a conic relationship. However, it is inherently non-convex due to the strict equality imposed on the terms. To address this challenge, a relaxation is introduced as recommended in [5]. The relaxed version is presented in (7.5):

$$\left\| \begin{matrix} 2z_{km} \\ z_{kk} - z_{mm} \end{matrix} \right\| \leq z_{kk} + z_{mm}. \tag{7.5}$$

Equation (7.5) transforms the non-convex equality constraint into a convex second-order conic constraint, making it suitable for SOCP-based optimization. This relaxation ensures computational efficiency while preserving the accuracy of the approximation within acceptable bounds [2].

## 7.2 Application to the OPF Problem

To obtain an equivalent second-order conic approximation for the exact multiperiod OPF model given by (2.28), the product of voltage variables is approximated using an auxiliary set of variables $z_{kk,h}$, $z_{km,h}$, and $z_{mm,h}$ including the conic condition defined in (7.4), which produces the second conic equivalent model defined in (7.6) [4].

$$\text{O.F.:} \quad \min\left\{E_{\text{loss}}\right\} = \sum_{k\in\mathbf{N}}\sum_{m\in\mathbf{N}}\sum_{h\in\mathbf{H}} g_{km} z_{km,h}\Delta_h \tag{7.6}$$

st:

$$\begin{aligned}
& p^{s}_{k,h} - p^{dg}_{k,h} - p^{d}_{k,h} = \sum_{m\in\mathbf{N}} g_{km} z_{km,h}, && \{\forall k \in \mathbf{N},\ \forall h \in \mathbf{H}\}\\
& {r_{km}}^2 j_{km,h} = z_{kk,h} - 2z_{km,h} + z_{mm,h}, && \{\forall km \in \mathbf{L},\ \forall h \in \mathbf{H}\}\\
& \left\| \begin{matrix} 2z_{km} \\ z_{kk} - z_{mm} \end{matrix} \right\| \le z_{kk} + z_{mm}, && \{\forall km \in \mathbf{L},\ \forall h \in \mathbf{H}\}\\
& \left| j_{km,h} \right| \le \left(i^{\max}_{km}\right)^2, && \{\forall km \in \mathbf{L},\ \forall h \in \mathbf{H}\}\\
& p^{s,\min}_{k,h} \le p^{s}_{k,h} \le p^{s,\max}_{k,h}, && \{\forall k \in \mathbf{N},\ \forall h \in \mathbf{H}\}\\
& p^{dg,\min}_{k,h} \le p^{dg}_{k,h} \le p^{dg,\max}_{k,h}, && \{\forall k \in \mathbf{N},\ \forall h \in \mathbf{H}\}\\
& \left(v^{\min}_{k}\right)^2 \le z_{kk,h} \le \left(v^{\max}_{k}\right)^2, && \{\forall k \in \mathbf{N},\ \forall h \in \mathbf{H}\}\\
& z_{kk,h} = \left(v_{\text{nom}}\right)^2, && \{\forall k \in \mathbf{S},\ \forall h \in \mathbf{H}\}.
\end{aligned}$$

It is worth mentioning that to obtain the optimization model (7.6), the constraints in the exact OPF model defined by (2.28) that depends on the voltages $v_{k,h}$ and $v_{m,h}$ must be transformed by using the new set of auxiliary variables $z_{kk,h}$, $z_{km,h}$, and $z_{mm,h}$, respectively. In the case of the current calculation per branch, the following algebraic manipulations are made:

$$r_{km} i_{km,h} = v_{k,h} - v_{m,h}, \qquad \{\forall km \in \mathbf{L},\ \forall h \in \mathbf{H}\}$$

$$r_{km}{}^2 i_{km,h}^2 = \left(v_{k,h} - v_{m,h}\right)^2, \qquad \{\forall km \in \mathbf{L},\ \forall h \in \mathbf{H}\}$$

$$r_{km}{}^2 i_{km,h}^2 = v_{k,h}^2 - 2v_{k,h}v_{m,h} + v_{m,h}^2, \qquad \{\forall km \in \mathbf{L},\ \forall h \in \mathbf{H}\}$$

$$r_{km}{}^2 j_{km,h} = z_{kk,h} - 2z_{km,h} + z_{mm,h}, \qquad \{\forall km \in \mathbf{L},\ \forall h \in \mathbf{H}\},$$

where is observed that an auxiliary variable called $j_{km,h}$ has been used to replace the square value of the current variable. The reader is invited to complete all the algebraic operations in the optimization model (2.28) in order to reach the conic approximation presented in model (7.6)

## 7.3 Numerical Validations

The validation of the second conic OPF formulation in (7.6) is made by comparing its results with the exact OPF model. The numerical implementations are made with the help of the `JuMP` optimization package and the `Ipopt` solver in the Julia programming language [6].

### 7.3.1 *DC Version of the 33-Bus System*

For this simulation, the DC version of the 33-bus grid is implemented [7]. This test feeder corresponds to a monopolar DC network operated with 12.66 kV in terminals of the substation located at bus 1. This system is composed of 33 nodes and 32 lines, i.e., it is a radial distribution grid [8]. The electrical configuration of the 33-bus grid is depicted in Fig. 7.1.

For this test system, the dispersed generation sources are located at nodes 12, 15, and 31, with sizes of 2400 kW each [9]. The electrical information regarding branch resistances and load consumptions is listed in Table 7.1.

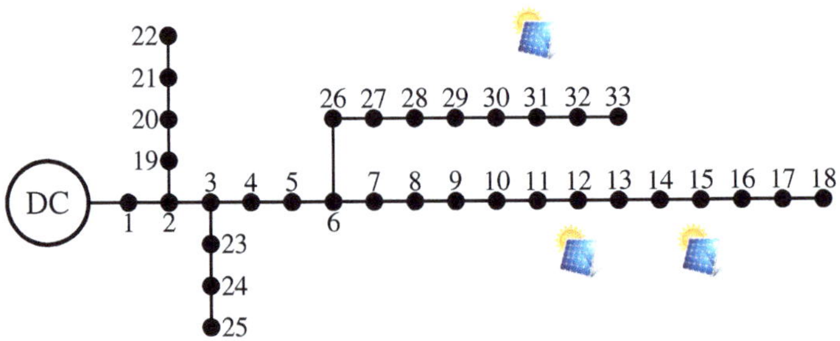


**Fig. 7.1** Schematic nodal connection of the monopolar DC version of the 33-bus grid

**Table 7.1** Network data regarding branches and loads for the 33-bus grid

| Line $l$ | Node $k$ | Node $m$ | $R_{km}$ (Ω) | $P_m$ (kW) | $I_l^{\max}$ (A) |
|---|---|---|---|---|---|
| 1 | 1 | 2 | 0.0922 | 100 | 320 |
| 2 | 2 | 3 | 0.4930 | 90 | 280 |
| 3 | 3 | 4 | 0.3660 | 120 | 195 |
| 4 | 4 | 5 | 0.3811 | 60 | 195 |
| 5 | 5 | 6 | 0.8190 | 60 | 195 |
| 6 | 6 | 7 | 0.1872 | 200 | 95 |
| 7 | 7 | 8 | 1.7114 | 200 | 85 |
| 8 | 8 | 9 | 1.0300 | 60 | 70 |
| 9 | 9 | 10 | 1.0400 | 60 | 55 |
| 10 | 10 | 11 | 0.1966 | 45 | 55 |
| 11 | 11 | 12 | 0.3744 | 60 | 55 |
| 12 | 12 | 13 | 1.4680 | 60 | 40 |
| 13 | 13 | 14 | 0.5416 | 120 | 40 |
| 14 | 14 | 15 | 0.5910 | 60 | 25 |
| 15 | 15 | 16 | 0.7463 | 60 | 20 |
| 16 | 16 | 17 | 1.2890 | 60 | 20 |
| 17 | 17 | 18 | 0.7320 | 90 | 20 |
| 18 | 2 | 19 | 0.1640 | 90 | 30 |
| 19 | 19 | 20 | 1.5042 | 90 | 25 |
| 20 | 20 | 21 | 0.4095 | 90 | 20 |
| 21 | 21 | 22 | 0.7089 | 90 | 20 |
| 22 | 3 | 23 | 0.4512 | 90 | 85 |
| 23 | 23 | 24 | 0.8980 | 420 | 70 |
| 24 | 24 | 25 | 0.8900 | 420 | 40 |
| 25 | 6 | 26 | 0.2030 | 60 | 85 |
| 26 | 26 | 27 | 0.2842 | 60 | 85 |
| 27 | 27 | 28 | 1.0590 | 60 | 70 |
| 28 | 28 | 29 | 0.8042 | 120 | 70 |
| 29 | 29 | 30 | 0.5075 | 200 | 55 |
| 30 | 30 | 31 | 0.9744 | 150 | 40 |
| 31 | 31 | 32 | 0.3105 | 210 | 25 |
| 32 | 32 | 33 | 0.3410 | 60 | 20 |

The expected daily generation profile in PV plants and the demand behavior for all the demand nodes are depicted in Fig. 7.2.

### 7.3.2 *Implementation of the SOCP Model for OPF Solution*

The first step for implementing the multiperiod OPF model (7.6) corresponds to upload all system parameters and generate the nodal conductance matrix, as presented below:

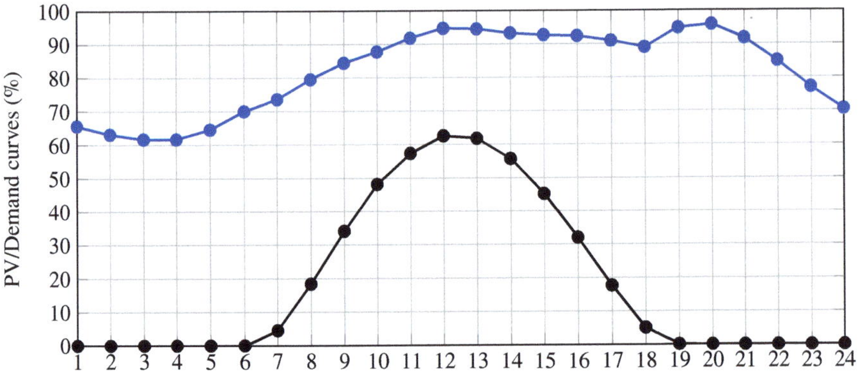


**Fig. 7.2** Expected demand and generation curves applied to the 33-bus grid

```
function Conductance(Lines,L,N)
    Gbus = zeros((N,N));
    for l = 1:L
        k = Int(Lines[l,1]);  m = Int(Lines[l,2]);
        rkm = Lines[l,3]; Gbus[k,m] = -1/rkm;
        Gbus[m,k] = -1/rkm;
        Gbus[k,k] = Gbus[k,k] + 1/rkm;
        Gbus[m,m] = Gbus[m,m] + 1/rkm;
    end
    return Gbus
end
using LinearAlgebra, DataFrames

Vb = 12660; Pb = 100; Rb = Vb^2/(Pb*1000); Ib = Pb*1000/Vb;
Lines =  DataFrames.DataFrame([
(1,  2,  0.0922, 320),(2,  3,  0.4930, 280),(3,  4,  0.3660, 195),
(4,  5,  0.3811, 195),(5,  6,  0.8190, 195),(6,  7,  0.1872, 95),
(7,  8,  1.7114, 85),(8,  9,  1.0300, 70), (9,  10, 1.0400, 55),
(10, 11, 0.1966, 55),(11, 12, 0.3744, 55), (12, 13, 1.4680, 40),
(13, 14, 0.5416, 40),(14, 15, 0.5910, 25), (15, 16, 0.7463, 20),
(16, 17, 1.2890, 20),(17, 18, 0.7320, 20), (2,  19, 0.1640, 30),
(19, 20, 1.5042, 25),(20, 21, 0.4095, 20), (21, 22, 0.7089, 20),
(3,  23, 0.4512, 85),(23, 24, 0.8980, 70), (24, 25, 0.8960, 40),
(6,  26, 0.2030, 85),(26, 27, 0.2842, 85), (27, 28, 1.0590, 70),
(28, 29, 0.8042, 70),(29, 30, 0.5075, 55), (30, 31, 0.9744, 40),
(31, 32, 0.3105, 25),(32, 33, 0.3410, 20),]);
DataFrames.rename!(Lines, [:k, :m, :Rkm, :Ikm])
Nodes = DataFrames.DataFrame([
(1,  1, 0,   0, 10),(2,  1, 100, 0, 0),(3,  1, 90,  0, 0),
(4,  1, 120, 0, 0),(5,  1, 60,  0, 0),(6,  1, 60,  0, 0),
(7,  1, 200, 0, 0),(8,  1, 200, 0, 0),(9,  1, 60,  0, 0),
(10, 1, 60,  0, 0),(11, 1, 45,  0, 0),(12, 1, 60,  0, 2400),
(13, 1, 60,  0, 0),(14, 1, 120, 0, 0),(15, 1, 60,  0, 2400),
(16, 1, 60,  0, 0),(17, 1, 60,  0, 0),(18, 1, 90,  0, 0),
```

```
(19, 1, 90,  0, 0),(20, 1, 90,  0, 0),(21, 1, 90,  0, 0),
(22, 1, 90,  0, 0),(23, 1, 90,  0, 0),(24, 1, 420, 0, 0),
(25, 1, 420, 0, 0),(26, 1, 60,  0, 0),(27, 1, 60,  0, 0),
(28, 1, 60,  0, 0),(29, 1, 120, 0, 0),(30, 1, 200, 0, 0),
(31, 1, 150, 0, 2400),(32, 1, 210, 0, 0),(33, 1, 60,  0, 0),])
DataFrames.rename!(Nodes, [:k, :Vk0, :Pk, :Pgdmin, :Pgdmax])
PVD = DataFrames.DataFrame([
(1,  0,           0.655092250),(2,  0,           0.630152498),
(3,  0,           0.615569653),(4,  0,           0.615829592),
(5,  0,           0.644566657),(6,  0,           0.698937446),
(7,  0.045411451, 0.734229507),(8,  0.184241504, 0.793483486),
(9,  0.340998605, 0.843313061),(10, 0.481610242, 0.876224102),
(11, 0.573748577, 0.917021370),(12, 0.625715388, 0.945950545),
(13, 0.618090573, 0.943876054),(14, 0.557162448, 0.931269100),
(15, 0.452363827, 0.925405805),(16, 0.320523656, 0.922597520),
(17, 0.176932056, 0.908069636),(18, 0.050658349, 0.888587275),
(19, 0.000498894, 0.946218065),(20, 0,           0.956175455),
(21, 0,           0.915546278),(22, 0,           0.847794326),
(23, 0,           0.768310994),(24, 0,           0.702968456),])
DataFrames.rename!(PVD, [:h, :PV, :Dm])
Lines.Rkm= Lines.Rkm/Rb;  Lines.Ikm = Lines.Ikm/Ib;
Nodes.Pk = Nodes.Pk/Pb; Nodes.Pgdmin = Nodes.Pgdmin/Pb;
Nodes.Pgdmax = Nodes.Pgdmax/Pb; vmin = 0.90; vmax = 1.10
```

The second step in the SOCP implementation corresponds to the construction of the OPF model defined in (7.6), as presented below:

```
using JuMP, Ipopt, Convex, CDCS, Clarabel, Gurobi
function SOCP(Nodes,Lines,PVD,vmin,vmax)
    N = size(Nodes,1); L = size(Lines,1); DT = size(PVD,1);
    Gbus = Conductance(Lines,L,N);
    elements = 1:N;  branches = 1:L; times = 1:DT;
    Pd = zeros(N,DT);
    Pgmin = zeros(N,DT); Pgmax = zeros(N,DT);
    Pvmin = zeros(N,DT);  Pvmax = zeros(N,DT);
    for h in times
        Pd[:,h]     = Nodes.Pk*PVD.Dm[h];
        Pvmax[:,h] = Nodes.Pgdmax*PVD.PV[h];
    end
    Pvmax[1,:] = zeros(1,DT);
    Pgmax[1,:] = Nodes.Pgdmax[1]*ones(1,DT);
    OPF = Model(Ipopt.Optimizer)
    #set_optimizer_attribute(OPF,"constr_viol_tol", 1e-15)
    @variable(OPF,pg[k in elements, h in times])
    @variable(OPF,pv[k in elements, h in times])
    @variable(OPF,z[k in elements, m in elements, h in times])
    @variable(OPF,j[l in branches, h in times])
    @objective(OPF,Min,sum(sum(sum(Gbus[k,m]*z[k,m,h]
    for k in elements) for m in elements) for h in times))
    for k = 1:N
        for h in times
            if k == 1
                @constraint(OPF,z[k,k,h] == 1);
```

```
            end
        @constraint(OPF,pg[k,h] >= Pgmin[k,h])
        @constraint(OPF,pg[k,h] <= Pgmax[k,h])
        @constraint(OPF,pv[k,h] >= Pvmin[k,h])
        @constraint(OPF,pv[k,h] <= Pvmax[k,h])
        for m = 1:N
        @constraint(OPF,z[k,m,h] <= vmax^2)
        @constraint(OPF,z[k,m,h] >= vmin^2)
        @constraint(OPF,(2*z[k,m,h])^2 + (z[k,k,h]-z[m,m,h])^2
        == (z[k,k,h]+z[m,m,h])^2)
        end
        @constraint(OPF, pg[k,h] + pv[k,h] - Pd[k,h] ==
        sum(Gbus[k,m]*z[k,m,h]  for m in elements))
        end
    end
    for l = 1:L
        for h in times
            k = Lines.k[l]; m = Lines.m[l]; rkm = Lines.Rkm[l];
            @constraint(OPF,j[l,h] ==
            (z[k,k,h] - 2*z[k,m,h] + z[m,m,h])/rkm^2)
            @constraint(OPF,0 <= j[l,h] <= Lines.Ikm[l]^2)
        end
    end
    optimize!(OPF)
   ploss = @show objective_value(OPF)*Pb;
   return j, pv, pg, ploss;
end
j, pv, pg, ploss = SOCP(Nodes,Lines,PVD,vmin,vmax);
global il = zeros(Float64, size(j, 1), 3)
for l = 1:size(j, 1)
    il[l, 1] = sqrt(maximum(value.(j[l, :]))) * Ib
    il[l, 2] = Lines.Ikm[l,1]*Ib;
    il[l, 3] = 100*il[l, 1]/il[l, 2];
end
ilmax= maximum(il[:,3]);
rilmax = argmax(il[:,3]);
```

where is observed that function `SOCP(Nodes,Lines,PVD,vmin,vmax)` corresponds to the structure that implements the optimization model (7.6) associated with the OPF solution in DC networks via the SOCP programming approximation.

### *7.3.3 Results Analysis and Discussion*

Once the Julia code is executed the following results are achieved.

- The final energy losses in this distribution network amount to 1224.8553 kWh/day, closely matching the result obtained from solving the exact OPF problem (1224.8548 kWh/day, see Ref. [10]). This minimal discrepancy highlights the

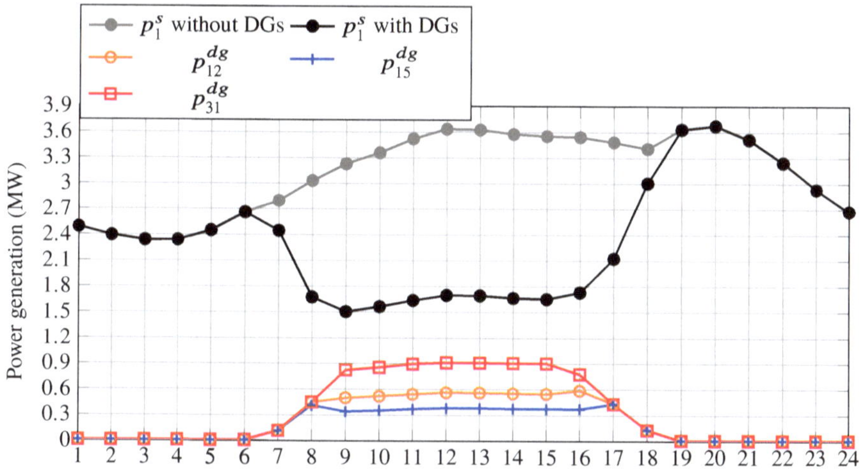


**Fig. 7.3** Daily power generation in conventional and renewable generation sources

effectiveness of the proposed SOCP model in achieving the OPF solution through a reliable and accurate convex approximation.

- The operation of the 33-bus DC network reports daily energy losses of about 2186.2803 kWh/day when distributed generators are disconnected. However, when these generators are operational, the energy losses are reduced by approximately 961.425 kWh/day, representing a significant improvement of 43.98% in this operational efficiency indicator.
- The maximum line loadability occurs on line 30, reaching 99.9933% of its maximum permissible current flow as defined by the thermal limit. Conversely, line 32 exhibits the lowest loadability, with a value of 23.8431% relative to its thermal capacity. These results confirm that the studied SOCP approximation effectively enforces thermal constraint limits across all distribution branches during every period of analysis, ensuring that the system operates within safe and reliable conditions.

On the other hand, Fig. 7.3 presents the daily power behavior in the slack node (conventional generator) and in all distributed generation sources.

The power behavior depicted in Fig. 7.3 allows observing the following facts:

- The power generation at the slack source, when distributed generators are connected, exhibits the same behavior as the initial operational case (without distributed generation) during nighttime periods, i.e., from hours 1 to 6 and 19 to 24. This behavior is expected, as there is no solar energy availability from distributed energy resources during these intervals.
- The renewable energy sources connected at nodes 12, 15, and 31 achieved maximum power outputs of 0.5775, 0.4246, and 0.9037 MW, respectively. This demonstrates that none of the distributed generators reached their maximum

allowable capacity of 1.4834 MW. These outcomes are primarily influenced by the thermal constraints imposed on the distribution line conductors, as well as the optimized coordination of all power sources, including the slack generator. This strategic coordination effectively contributes to minimizing the daily energy losses, as evidenced by the results, confirming the reports presented by the authors of [10].

## 7.4 Conclusions

The application of SOCP to solve the OPF problem in DC distribution networks has demonstrated its effectiveness in addressing nonlinear constraints through convex reformulations. By employing auxiliary variables and conic relaxations, the proposed methodology ensures computational efficiency while maintaining high accuracy in the solutions. Numerical validations on the 33-bus DC test system highlighted significant improvements in operational efficiency, with energy losses reduced by 43.98% when distributed generation units were incorporated. Additionally, the results confirm that the thermal and operational constraints across all branches were respected, guaranteeing safe and reliable grid operation. The optimized coordination between renewable energy sources and the slack generator effectively minimized daily energy losses, even when DG units operated below their maximum capacities due to thermal constraints. These findings underline the robustness and scalability of the SOCP-based approach, making it a valuable tool for enhancing the performance of modern DC distribution networks with renewable energy integration.

## References

1. Boyd, S., & Vandenberghe, L. (2004). *Convex optimization*. Cambridge University Press.
2. Gan, L., & Low, S. H. (2014). Optimal power flow in direct current networks. *IEEE Transactions on Power Systems, 29*(6), 2892–2904.
3. Chen, Y., Xiang, J., & Li, Y. (2018). SOCP relaxations of optimal power flow problem considering current margins in radial networks. *Energies, 11*(11), 3164.
4. Montoya, O. D., Arias-Londoño, A., & Molina-Cabrera, A. (2022). Branch optimal power flow model for DC networks with radial structure: A conic relaxation. *Tecnura, 26*(71), 30–42.
5. Farivar, M., & Low, S. H. (2013). Branch flow model: Relaxations and convexification—part I. *IEEE Transactions on Power Systems, 28*(3), 2554–2564.
6. Lubin, M., Dowson, O., Dias Garcia, J., Huchette, J., Legat, B., & Vielma, J. P. (2023). JuMP 1.0: Recent improvements to a modeling language for mathematical optimization. *Mathematical Programming Computation*.
7. Montoya, O. D., Gil-González, W., & Garces, A. (2020). Numerical methods for power flow analysis in DC networks: State of the art, methods and challenges. *International Journal of Electrical Power & Energy Systems, 123*, 106299.

8. Guzmán-Henao, J. A., Cortés-Caicedo, B., Restrepo-Cuestas, B. J., Bolaños, R. I., & Grisales-Noreña, L. F. (2024). Optimal integration of photovoltaic generators into urban and rural power distribution systems. *Solar Energy, 270*, 112400.
9. Montoya, O. D., Grisales-Noreña, L. F., & Giral-Ramírez, D. A. (2023). Multi-objective dispatch of PV plants in monopolar DC grids using a weighted-based iterative convex solution methodology. *Energies, 16*(2), 976.
10. Montoya, O. D., Grisales-Noreña, L. F., & Hernández, J. C. (2023). Efficient day-ahead dispatch of photovoltaic sources in monopolar DC networks via an iterative convex approximation. *Energies, 16*(3), 1105.

# Chapter 8
# Third Conic Approximation

**Abstract** This chapter presents a second-order cone programming (SOCP) approximation for solving the optimal power flow (OPF) problem in monopolar DC networks. The proposed approach transforms the original non-convex problem into a convex formulation by leveraging the properties of second-order cones, enabling efficient and scalable optimization. The methodology is validated using numerical simulations implemented in Julia, employing the `JuMP` package and solvers such as Ipopt and Gurobi. The results demonstrate that the SOCP model closely approximates the exact OPF solution, with minimal error differences. Additionally, the impact of the solver choice on accuracy and computational efficiency is analyzed, highlighting the robustness of the proposed method in power flow optimization. The findings confirm that SOCP provides an effective and computationally efficient alternative for OPF analysis in DC networks.

**Keywords** Second-order cone programming · Optimal power flow · DC networks · Convex reformulation · Numerical simulation

## 8.1 Second-Order Cone Programming Theory

Second-order cone (SOC) programming theory is based on the use of constraints that define a convex set in the shape of a cone [1, 2]. These cones, known as second-order cones or quadratic cones, are characterized by their geometric structure that facilitates the solution of complex optimization problems [3].

A second-order cone is mathematically defined as

$$\mathcal{K} = \{x \in \mathbb{R}^n : x_0 \geq \|x_1\|_2\}, \tag{8.1}$$

where $x = (x_0, x_1) \in \mathbb{R}^n$ is a vector, $x_0$ is the scalar component (also called the "height" of the cone); $x_1 \in \mathbb{R}^{n-1}$ is a vector in $(n-1)$-dimensions, and $\|x_1\|_2$ represents the Euclidean norm of the vector $x_1$.

Geometrically, the second-order cone can be visualized as a circular cone in a higher-dimensional space, where the base of the cone is a sphere and the axis

© The Author(s), under exclusive license to Springer Nature Switzerland AG 2026

O. D. Montoya Giraldo et al., *Optimal Power Flow Methods for DC Networks*, Power Systems, https://doi.org/10.1007/978-3-032-05734-1_8

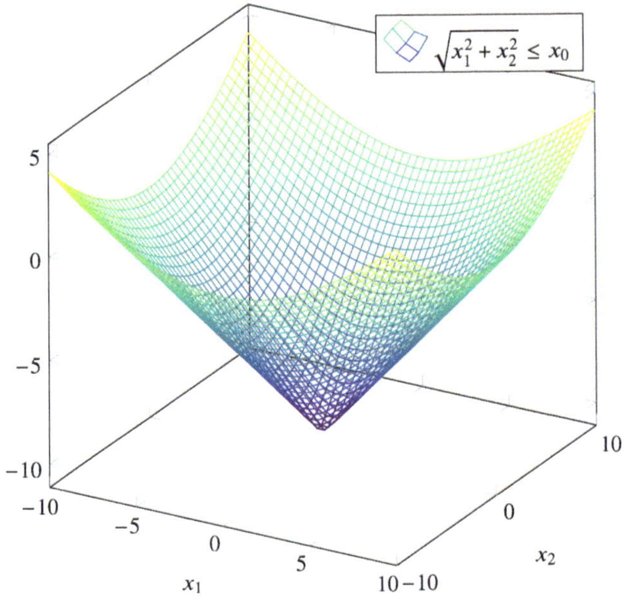


**Fig. 8.1** Lorentz cone in the three-dimensional space with $x_0 = 10$

corresponds to $x_0$. To give an example in the three-dimensional space, it is used the general structure of the Lorentz cone which is depicted in Fig. 8.1 with $x_0 = 10$.

In the context of optimization, constraints based on second-order cones allow modeling problems with quadratic terms, such as power limits, currents, or energy, that would otherwise be difficult to handle with linear programming techniques.

The second-order cone has several properties that make it useful for optimization [3, 4]: (i) It is a convex set, which guarantees the existence of globally optimal solutions for well-formulated problems. (ii) Cone-shaped constraints can represent nonlinear relationships, such as quadratic inequalities. (iii) Its structure lends itself to being solved by efficient convex optimization algorithms, such as the interior-point method.

## 8.2 General SOC Approximation

SOC programming approximation is a technique that allows non-convex problems to be transformed into convex problems that can be solved efficiently. This is especially useful in applications such as power systems, structural design, and signal processing [5].

Suppose we have a typical non-convex quadratic problem, formulated as

$$\min \quad z = \frac{1}{2} x^\top Q x + c^\top x,$$

$$\text{subject to:} \tag{8.2}$$

$$\|A_i x + b_i\|_2^2 \leq (d_i^\top x + e_i)^2, \quad \forall i,$$

where $z$ represents the objective function, $A \in \mathbb{R}^{n \times n}$ is a square cost matrix, $C \in \mathbb{R}^n$ is a column vector, and $x \in \mathbb{R}^n$ is the decision variable vector. $A \in \mathbb{R}^{m \times n}$ is a rectangular matrix, $b \in \mathbb{R}^m$ is a vector, $d_i \in \mathbb{R}^n$ is a column vector, and $e_i$ is a scalar.

This problem can be transformed into an SOC format by applying the property of second-order cones. We rewrite the quadratic constraint as

$$\min \quad z = \frac{1}{2} x^\top Q x + C^\top x,$$

$$\text{subject to:} \tag{8.3}$$

$$\|A_i x + b_i\|_2 \leq d_i^\top x + e_i,$$

which defines a convex constraint compatible with the SOCP model.

Transforming to SOC format offers several benefits [6, 7]: (i) converts complex constraints into more manageable convex constraints, (ii) allows the use of specialized solvers such as Mosek, Gurobi, or SCS to solve large-scale problems, and (iii) reduces sensitivity to numerical problems by working with well-defined convex constraints.

It is important to mention that although SOCP is a powerful technique, not all problems can be directly transformed into this format. Some non-convex constraints may require additional approximations or simplifications that could introduce errors into the final solution.

## 8.3 Application to the OPF Problem

To achieve an equivalent SOC programming for the exact multiperiod OPF model given by (2.28), first it is necessary to set out the power flow equation and then transform the product of the set of variables into an SOC constraint. For this purpose, it is defined the power flow $p_{km,h}$ in a branch $km$ similarly as was defined the current flow in a branch in (2.19) [8]. The power flow $p_{km,h}$ can be calculated as

$$p_{km,h} = v_{k,h} y_{km} \left( v_{k,h} - v_{m,h} \right), \ \{\forall km \in \mathbf{L}, \ \forall h \in \mathbf{H}\} \tag{8.4}$$

$$p_{mk,h} = v_{m,h} y_{km} \left( v_{m,h} - v_{m,h} \right), \ \{\forall km \in \mathbf{L}, \ \forall h \in \mathbf{H}\}, \tag{8.5}$$

where $p_{km,h}$ is the power flowing from node $k$ to node $k$, which is commonly called sending power flow, while $p_{mk,h}$ is the power flowing from node $m$ to node $k$, which is called receiving power flow. These powers are not the same, i.e., $p_{km,h} \neq p_{mk,h}$; $y_{km} = \frac{1}{r_{km}}$ is the admittance associated with the line used to interconnect the nodes $k$ and $m$. $\mathbf{L}$ and $\mathbf{H}$ are the sets that contain all the branches of the monopolar DC network and all periods under analysis.

The power flow loss in branch $km$ is computed as the sum of the sending and receiving power flows in line $km$, as follows:

$$\begin{aligned} p^{loss}_{km,h} =& p_{km,h} + p_{mk,h} = \\ =& y_{km}\left(v^2_{k,h} - 2v_{k,h}v_{m,h} + v^2_{m,h}\right), \ \{\forall km \in \mathbf{L},\ \forall h \in \mathbf{H}\}. \end{aligned} \tag{8.6}$$

The power flow constraints (8.4) and (8.5), along with the power loss equation (8.6), are inherently nonlinear and non-convex. This characteristic makes finding a global optimum for the overall problem challenging and computationally intensive. Nonetheless, these constraints can be reformulated as linear equations by introducing two auxiliary variables, as described below:

$$u_{k,h} = v^2_{k,t}, \ \{\forall k \in \mathbf{N},\ \forall h \in \mathbf{H}\} \tag{8.7}$$

$$w_{km,t} = v_{k,t}v_{m,t}, \ \{\forall k \in \mathbf{N},\ \forall h \in \mathbf{H}\}, \tag{8.8}$$

where $\mathbf{N}$ is the set that contains all the nodes of the monopolar DC network.

By replacing these auxiliary variables into (8.4), (8.5), and the power loss expression (8.6), the resulting formulations are expressed as follows:

$$p_{km,h} = y_{km}\left(u_{k,h} - w_{km,h}\right), \tag{8.9}$$

$$p_{mk,h} = y_{km}\left(u_{m,h} - w_{km,h}\right), \tag{8.10}$$

$$p^{loss}_{km,h} = y_{km}\left(u_{k,h} - 2w_{km,h} + u_{m,h}\right). \tag{8.11}$$

Although the constraints (8.9), (8.10), and (8.11) are linear, the auxiliary variables defined in (8.7) and (8.8) are continuous being non-convex constraints. Hence it is necessary to construct a convex model, which is defined as follows:

$$\begin{aligned} w_{km,h} =& v_{k,h}v_{m,h} \\ w_{km,h}w_{km,h} =& v^2_{k,h}v^2_{m,h} = \left\| w_{km,h} \right\|^2 = u_{k,h}u_{m,h} \\ \left\| \begin{matrix} 2w_{km,h} \\ u_{k,h} - u_{m,h} \end{matrix} \right\| =& u_{k,h} + u_{m,h} \\ \left\| \begin{matrix} 2w_{km,h} \\ u_{k,h} - u_{m,h} \end{matrix} \right\| \leq& u_{k,h} + u_{m,h}. \end{aligned} \tag{8.12}$$

Lastly, the constraint representing the power flow balance at node $k$ can be expressed as

$$p_{k,h}^{s} + p_{k,h}^{dg} - p_{k,h}^{d} = \sum_{km \in L} \left(A_{kl}^{+} p_{km,h} + A_{kl}^{-} p_{mk,h}\right). \tag{8.13}$$

Here $A = A^{+} + A^{-}$ represents the incidence matrix, where $A^{+}$ includes the positive entries corresponding to outgoing flows from each node, and $A^{-}$ contains the negative entries corresponding to incoming flows.

Model (8.14) models the SOC programming approximation to deal with the OPF problem in monopolar DC networks.

**SOC Multiperiod OPF Approximation**

$$\text{O.F.:} \quad \min\left\{E_{\text{loss}}\right\} = \sum_{h \in \mathbf{H}} \sum_{k \in \mathbf{N}} \sum_{m \in \mathbf{N}} p_{km,h}^{loss}, \tag{8.14}$$

st:

$$\begin{aligned}
& p_{k,h}^{s} + p_{k,h}^{dg} - p_{k,h}^{d} = \sum_{km \in L} \left(A_{kl}^{+} p_{km,h} + A_{kl}^{-} p_{mk,h}\right), && \{\forall k \in \mathbf{N},\ \forall h \in \mathbf{H}\} \\
& p_{km,h} = y_{km}\left(u_{k,h} - w_{km,h}\right), && \{\forall km \in \mathbf{L},\ \forall h \in \mathbf{H}\} \\
& p_{mk,h} = y_{km}\left(u_{m,h} - w_{km,h}\right), && \{\forall km \in \mathbf{L},\ \forall h \in \mathbf{H}\} \\
& \left|p_{km,h}\right| \leq p_{km}^{\max}, && \{\forall km \in \mathbf{L},\ \forall h \in \mathbf{H}\} \\
& p_{k,h}^{s,\min} \leq p_{k,h}^{s} \leq p_{k,h}^{s,\max}, && \{\forall k \in \mathbf{N},\ \forall h \in \mathbf{H}\} \\
& p_{k,h}^{dg,\min} \leq p_{k,h}^{dg} \leq p_{k,h}^{dg,\max}, && \{\forall k \in \mathbf{N},\ \forall h \in \mathbf{H}\} \\
& \left(v_{k}^{\min}\right)^{2} \leq u_{k,h} \leq \left(v_{k}^{\max}\right)^{2}, && \{\forall k \in \mathbf{N},\ \forall h \in \mathbf{H}\} \\
& \left\| \begin{matrix} 2w_{km,h} \\ u_{k,h} - u_{m,h} \end{matrix} \right\| \leq u_{k,h} + u_{m,h}, && \left\{\begin{matrix} \forall km \in \mathbf{L},\ \forall h \in \mathbf{H} \\ \forall k \in \mathbf{N} \end{matrix}\right\} \\
& u_{k,h} = v_{\text{nom}}^{2}, && \{\forall k \in \mathbf{S},\ \forall h \in \mathbf{H}\}.
\end{aligned}$$

## 8.4 Numerical Validations

The validation of the SOC multiperiod OPF approximation presented in (8.14) is carried out using the same test system described in Chap. 7. The numerical implementations are performed with the help of the `JuMP` optimization package and the `Ipopt` solver in the Julia programming language [9].

### 8.4.1 Implementation of SOCP Model for OPF Solution

The implementation of the SOCP model (8.14) requires to upload all system parameters to generate the positive and negative incidence matrices ($A^+$ and $A^-$) and admittance vector of the branches ($y_{km}$), as presented below:

```
function IncidenceMatrices(Lines,L,N)
    Apos = zeros((N, L))
    Aneg = zeros((N, L))
    for k in 1:L
        ni =  Lines.k[k];
        nj =  Lines.m[k];
        Apos[ni, k] = 1
        Aneg[nj, k] = 1
    end
    return Apos, Aneg
end

using LinearAlgebra, DataFrames

Vb = 12660; Pb = 100; Rb = Vb^2/(Pb*1000); Ib = Pb*1000/Vb;
Lines =  DataFrames.DataFrame([
(1,  2,  0.0922, 320),(2,  3,  0.4930, 280),(3,  4,  0.3660, 195),
(4,  5,  0.3811, 195),(5,  6,  0.8190, 195),(6,  7,  0.1872, 95),
(7,  8,  1.7114, 85),(8,  9,  1.0300, 70), (9,  10, 1.0400, 55),
(10, 11, 0.1966, 55),(11, 12, 0.3744, 55), (12, 13, 1.4680, 40),
(13, 14, 0.5416, 40),(14, 15, 0.5910, 25), (15, 16, 0.7463, 20),
(16, 17, 1.2890, 20),(17, 18, 0.7320, 20), (2,  19, 0.1640, 30),
(19, 20, 1.5042, 25),(20, 21, 0.4095, 20), (21, 22, 0.7089, 20),
(3,  23, 0.4512, 85),(23, 24, 0.8980, 70), (24, 25, 0.8960, 40),
(6,  26, 0.2030, 85),(26, 27, 0.2842, 85), (27, 28, 1.0590, 70),
(28, 29, 0.8042, 70),(29, 30, 0.5075, 55), (30, 31, 0.9744, 40),
(31, 32, 0.3105, 25),(32, 33, 0.3410, 20),]);
DataFrames.rename!(Lines, [:k, :m, :Rkm, :Ikm])
Nodes = DataFrames.DataFrame([
(1,  1, 0,   0, 10),(2,  1, 100, 0, 0),(3,  1, 90,  0, 0),
(4,  1, 120, 0, 0),(5,  1, 60,  0, 0),(6,  1, 60,  0, 0),
(7,  1, 200, 0, 0),(8,  1, 200, 0, 0),(9,  1, 60,  0, 0),
(10, 1, 60,  0, 0),(11, 1, 45,  0, 0),(12, 1, 60,  0, 2400),
(13, 1, 60,  0, 0),(14, 1, 120, 0, 0),(15, 1, 60,  0, 2400),
(16, 1, 60,  0, 0),(17, 1, 60,  0, 0),(18, 1, 90,  0, 0),
(19, 1, 90,  0, 0),(20, 1, 90,  0, 0),(21, 1, 90,  0, 0),
(22, 1, 90,  0, 0),(23, 1, 90,  0, 0),(24, 1, 420, 0, 0),
(25, 1, 420, 0, 0),(26, 1, 60,  0, 0),(27, 1, 60,  0, 0),
(28, 1, 60,  0, 0),(29, 1, 120, 0, 0),(30, 1, 200, 0, 0),
(31, 1, 150, 0, 2400),(32, 1, 210, 0, 0),(33, 1, 60,  0, 0),])
DataFrames.rename!(Nodes, [:k, :Vk0, :Pk, :Pgdmin, :Pgdmax])
PVD = DataFrames.DataFrame([
(1,  0,            0.655092250),(2,  0,            0.630152498),
(3,  0,            0.615569653),(4,  0,            0.615829592),
(5,  0,            0.644566657),(6,  0,            0.698937446),
(7,  0.045411451,  0.734229507),(8,  0.184241504,  0.793483486),
(9,  0.340998605,  0.843313061),(10, 0.481610242,  0.876224102),
```

```
(11, 0.573748577,  0.917021370),(12, 0.625715388,  0.945950545),
(13, 0.618090573,  0.943876054),(14, 0.557162448,  0.931269100),
(15, 0.452363827,  0.925405805),(16, 0.320523656,  0.922597520),
(17, 0.176932056,  0.908069636),(18, 0.050658349,  0.888587275),
(19, 0.000498894,  0.946218065),(20, 0,            0.956175455),
(21, 0,            0.915546278),(22, 0,            0.847794326),
(23, 0,            0.768310994),(24, 0,            0.702968456),])
DataFrames.rename!(PVD, [:h, :PV, :Dm])
Lines.Rkm= Lines.Rkm/Rb;  Lines.Ikm = Lines.Ikm/Ib;
Nodes.Pk = Nodes.Pk/Pb; Nodes.Pgdmin = Nodes.Pgdmin/Pb;
Nodes.Pgdmax = Nodes.Pgdmax/Pb; vmin = 0.90; vmax = 1.10;
```

The next step is to implement the SOCP model presented in (8.14), as follows:

```
using JuMP, Ipopt
function solveSOC(Nodes,Lines,PVD,vmin,vmax)
    N = size(Nodes,1); L = size(Lines,1); DT = size(PVD,1);
    Apos, Aneg =  IncidenceMatrices(Lines,L,N);
    elements = 1:N;  branches = 1:L; times = 1:DT;
    Pd = zeros(N,DT);
    posPV = findall(x -> x != 0, Nodes.Pgdmax)
    NumPV = length(posPV)
    Pvmax = zeros(NumPV,DT);
    for h in times
        Pd[:,h] = Nodes.Pk*PVD.Dm[h];
        for k = 1:NumPV
            Pvmax[k,h] = Nodes.Pgdmax[posPV[k]]*PVD.PV[h];
        end
    end
    OPF = Model(Ipopt.Optimizer)

    @variable(OPF,pg[1, h in times]>= 0 )
    @variable(OPF,pv[1:NumPV, h in times] >= 0)
    @variable(OPF,pkm[l in branches, h in times])
    @variable(OPF,pmk[l in branches, h in times])
    @variable(OPF,uk[k in elements, h in times])
    @variable(OPF,wkm[l in branches, h in times])
    @objective(OPF,Min,sum(sum(pkm[l,h]+ pmk[l,h]
    for l in branches) for h in times))
    for h in times
        for k = 1:N
            if k == 1
                @constraint(OPF,uk[k,h] == 1);
                @constraint(OPF,pg[k,h] - Pd[k,h] ==
                dot(Apos[k, :], Vector(pkm[:, h])) +
                dot(Aneg[k, :], Vector(pmk[:, h])))
            elseif any(k -> k in posPV, [k])
                pos = findall(x -> x == k, posPV)[1]
                @constraint(OPF,pv[pos,h] - Pd[k,h] ==
                dot(Apos[k, :], Vector(pkm[:, h])) +
                dot(Aneg[k, :], Vector(pmk[:, h])))
                @constraint(OPF,pv[pos,h] <= Pvmax[pos,h])
            else
```

```
                @constraint(OPF, - Pd[k,h] == dot(Apos[k, :],
                Vector(pkm[:, h])) + dot(Aneg[k, :],
                Vector(pmk[:, h])))
            end
            @constraint(OPF,uk[k,h] <= vmax^2)
            @constraint(OPF,uk[k,h] >= vmin^2)
        end
        for l in 1:L
            k = Lines.k[l]; m = Lines.m[l]; ykm = 1/Lines.Rkm[l];
            @constraint(OPF,pkm[l,h] == (uk[k,h]-wkm[l,h])*ykm)
            @constraint(OPF,pmk[l,h] == (uk[m,h]-wkm[l,h])*ykm)
            @constraint(OPF,(2*wkm[l,h])^2 + (uk[k,h]-uk[m,h])^2
            <= (uk[k,h] + uk[m,h])^2)
            @constraint(OPF,-Lines.Ikm[l]<=pkm[l,h]<=Lines.Ikm[l])
            @constraint(OPF,-Lines.Ikm[l]<=pmk[l,h]<=Lines.Ikm[l])
            @constraint(OPF, pkm[l, h] + pmk[l, h] >= 0)

        end
    end
    optimize!(OPF)
    ploss = @show objective_value(OPF)*Pb/1000;
    vol = JuMP.value.(uk)
    for h in times
        for k in elements
            vol[k,h] = sqrt(vol[k,h])
        end
    end
    pv = JuMP.value.(pv)
    pg = JuMP.value.(pg)
    return vol, pv, pg, ploss;
end

vol, pv, pg, ploss = solveSOC(Nodes,Lines,PVD,vmin,vmax);
```

Note that the function `solveSOC(Nodes,Lines,PVD,vmin,vmax)` executes the SOCP model presented in (8.14) where the outputs of the function are the final energy losses, powers, the generator power, photovoltaic powers, and nodal voltages.

### 8.4.2 Results Analysis and Discussion

Once the Julia code is run the following results are obtained:

- The energy losses reached using the SOCP model (8.14) amount to 1224.0612kWh/day. This result is close to its solution obtained with the exact OPF problem which is 1224.8548 kWh/day [10]. This difference between the two results is minimal, which is around 0.064%. This demonstrates the effectiveness of the proposed SOCP model. However, this result also depends on the solver used.

- The energy losses obtained using the Gurobi solver in the SOCP model (8.14) amount to 1224.8621 kWh/day. This result is even closer to the exact solution, with an error of approximately 0.0006%. This shows that the choice of solver significantly influences the accuracy of the SOCP model's results. While the approximation errors remain minimal, selecting an appropriate solver can further reduce discrepancies and enhance computational efficiency. These findings highlight the robustness of the SOCP model in approximating the exact OPF solution while demonstrating that solver optimization plays a crucial role in achieving high-precision results.

## 8.5 Conclusions

The application of SOCP to solve the OPF problem in monopolar DC networks has proven to be an effective approach for handling nonlinear constraints through convex reformulations. By introducing auxiliary variables and leveraging conic relaxations, the proposed methodology ensures computational efficiency while maintaining high accuracy in the solutions. Numerical validations confirmed that the SOCP approximation closely matches the results of the exact OPF formulation, with an error margin below 0.064%. Additionally, the choice of solver significantly impacts accuracy, as demonstrated by the reduced approximation error of 0.0006% when using the Gurobi solver. These findings highlight the robustness of SOCP in optimizing power flow while ensuring compliance with operational constraints. The approach provides a scalable and computationally efficient solution for DC network optimization, making it a valuable tool for modern power system applications.

## References

1. Le Thi, H. A., Le, H. M., & Dinh, T. P. (2019). *Optimization of complex systems: Theory, models, algorithms and applications* (vol. 991). Springer.
2. Garcés, A. (2021). *Mathematical programming for power systems operation with applications in Python: From theory to applications in Python*. Wiley.
3. Boyd, S., & Vandenberghe, L. (2004). *Convex optimization*. Cambridge University Press.
4. Bonnans, F., & Deletaille, A. L. (2019). *Convex and stochastic optimization*. Universitext. Springer.
5. Vandenberghe, L., & Boyd, S. (1998). Applications of second-order cone programming. *Linear Algebra and its Applications, 284*, 343–374.
6. Alzalg, B., & Alioui, H. (2021). Applications of stochastic mixed-integer second-order cone optimization. *IEEE Access, 10*, 3522–3547.
7. Useche-Arteaga, M., Gil-González, W., Gomis-Bellmunt, O., Cheah-Mane, M., & Lacerda, V. (2025). Robust energy management in active distribution networks using mixed-integer convex optimization. *Electric Power Systems Research, 241*, 111367.
8. Gil-González, W., Montoya, O. D., & Hernández, J. C. (2023). An energy management system for the optimal operation of BESS in DC microgrids: A robust convex programming approach. *IEEE Access, 11*, 38168–38181.

9. Lubin, M., Dowson, O., Dias Garcia, J., Huchette, J., Legat, B., & Vielma, J. P. (2023). JuMP 1.0: Recent improvements to a modeling language for mathematical optimization. *Mathematical Programming Computation*, .
10. Montoya, O. D., Grisales-Noreña, L. F., & Hernández, J. C. (2023). Efficient day-ahead dispatch of photovoltaic sources in monopolar DC networks via an iterative convex approximation. *Energies, 16*(3), 1105.

# Chapter 9
# Quasi-static Time-Series Optimal Power Flow in DC Distribution Networks with Energy Storage

**Abstract** A quasi-static optimal power flow is a type of analysis that incorporates time-varying system conditions, such as load and generation variations, without a full dynamic simulation. Energy storage plays a key role in this analysis since it couples the problem in the time. In addition to the usual nonlinearity difficulties associated with load flow, this problem must deal with the fact that the optimization problem is large-scale and time-varying. This chapter presents an algorithm based on dual-decomposition to solve this challenging problem.

**Keywords** Quasi-static optimal power flow · DC distribution networks · Energy storage · Time-series analysis · Dual decomposition

## 9.1 Quasi-Static Optimal Power Flow

The optimal power flow problem requires to be adapted to consider the variation of load and generation in a time window. This methodology constitutes the quasi-static optimal power flow. It consists of iteratively solving a series of sequential instances of the optimal power flow problem. Therefore, the computational complexity is increased, generating a large-scale optimization problem. Despite this challenge, the structure of the model allows for distributed algorithms such as the dual-decomposition and the alternating direction method of multipliers [1]. Energy storage plays a key role in this problem since it couples the problem in the time.

The terms quasi-dynamic and quasi-static are often interchangeable [2]. Both refer to simulations that are in between transient behavior and steady state [3]. This type of analysis is required for screening the impact of distributed energy resources in a time-series, an aspect that static power flow analysis does not capture. Quasi-static analyses are becoming popular in power distribution applications [4]. Even free software products have been developed to simplify this type of analysis for conventional power distribution [5]. However, the problem has been little studied in DC distribution networks.

The grid is represented by an oriented graph $\mathcal{G} = \{\mathcal{N}, \mathcal{E}\}$, where $\mathcal{N}$ is the set of nodes and $\mathcal{E} \subseteq \mathcal{N} \times \mathcal{N}$ the set of edges (lines). We assume there is only one line

© The Author(s), under exclusive license to Springer Nature Switzerland AG 2026

O. D. Montoya Giraldo et al., *Optimal Power Flow Methods for DC Networks*,
Power Systems, https://doi.org/10.1007/978-3-032-05734-1_9

per edge. The nodal admittance matrix is $G = [g_{km}]$. Demanded power is given by $d = [d_{kt}]$. Nodal variables include generated power $p = [p_{kt}]$, stored power $s = [s_{kt}]$, stored energy $e = [e_{kt}]$, and terminal voltage $v = [v_{kt}]$, where $k \in \mathcal{N}$ represents the node and $t$ the time. The cost of the energy for each time is given by $c_t$. Maximum values are represented by variables with overline and minimum values by underline. Thus, the quasi-static optimal power flow can be represented as follows:

$$\begin{aligned}
\min \sum_{t\in\mathcal{T}} \ell_t(v), \\
v_{1t} &= 1, \\
p_{kt} - s_{kt} - d_{kt} &= \sum_{m\in\mathcal{N}} g_{km} v_{kt} v_{mt}, \ \forall k \in \mathcal{N}, \forall t \in \mathcal{T}, \\
\underline{v}_{kt} \leq v_{kt} &\leq \overline{v}_{kt}, \forall k \in \mathcal{N}, \forall t \in \mathcal{T}, \\
\underline{p}_{kt} \leq p_{kt} &\leq \overline{p}_{kt}, \forall k \in \mathcal{N}, \forall t \in \mathcal{T}, \\
-\overline{s}_{kt} \leq s_{kt} &\leq \overline{s}_{kt}, \forall k \in \mathcal{N}, \forall t \in \mathcal{T}, \\
0 \leq e_{kt} &\leq \overline{e}_{kt}, \ \forall k \in \mathcal{N}, \forall t \in \mathcal{T}, \\
-\overline{p}_{km} \leq g_{km} v_{kt}(v_{kt} - v_{mt}) &\leq \overline{p}_{km}, \ \forall km \in \mathcal{E}, \forall t \in \mathcal{T}, \\
e_{kt+1} &= e_{kt} + s_{kt}, \ \forall k \in \mathcal{N}, \forall t \in \mathcal{T}, \\
e_{k1} &= 0.
\end{aligned} \tag{9.1}$$

In this model, the objective function can be loss or cost minimization. Constraints include the power balance, maximum and minimum limits for each decision variable, and energy balance on the storage devices. Time is represented by $t \in \mathcal{T}$, where $\mathcal{T}$ is the horizon to be analyzed. For loss minimization, the objective function is given by the following quadratic form:[1]

$$\ell_t(v) = c_t \sum_{k\in\mathcal{N}} \sum_{m\in\mathcal{N}} g_{km} v_{kt} v_{mt}, \tag{9.2}$$

which is a convex-quadratic function due to the fact that $G$ is positive semidefinite. Other objectives such as cost minimization or optimal power-sharing can be included into this objective.

In the interest of brevity, when clear from the context, we assume that $k \in \mathcal{N}$, $km \in \mathcal{E}$, and $t \in \mathcal{T}$. Moreover, we spit the variables and the constraints into two groups, namely $\{v, p\} \in \Omega_t$ and $\{s, e\} \in \Gamma_t$. With this convenient representation,

[1] For cost minimization the objective function is linear with respect to $p$, but the rest of the development is the same.

we can rewrite model (9.1) as presented below:

$$\begin{aligned} \min \sum_t \ell_t(v) & \\ \{v, p\} \in \Omega_t, & \\ \{s, e\} \in \Gamma_t, & \\ h_{kt}(v, p) = s_{kt}, & \end{aligned} \tag{9.3}$$

where $\Omega_t$ and $\Gamma_t$ are sets given by

$$\Omega_t = \left\{ \begin{array}{c} v_{1t} = 1 \\ \underline{v}_{kt} \leq v_{kt} \leq \overline{v}_{kt} \\ \underline{p}_{kt} \leq p_{kt} \leq \overline{p}_{kt} \\ -\overline{p}_{km} \leq g_{km} v_{kt}(v_{kt} - v_{mt}) \leq \overline{p}_{km} \end{array} \right\}, \tag{9.4}$$

and

$$\Gamma_t = \left\{ \begin{array}{c} -\overline{s}_{kt} \leq s_{kt} \leq \overline{s}_{kt} \\ 0 \leq e_{kt} \leq \overline{e}_{kt} \\ e_{kt+1} = e_{kt} + s_{kt} \\ e_{1t} = 0 \end{array} \right\}. \tag{9.5}$$

Besides, $h_{kt}$ are functions that couple the two sets, namely

$$h_{kt}(v, p) = p_{kt} - d_{kt} - \sum_m g_{km} v_{kt} v_{mt}. \tag{9.6}$$

This constraint is non-convex though it can be convexified via linearization or second-order cone approximations. Notice that (9.3) is completely equivalent to (9.1). It is just a convenient representation that reveals a structure that may be used to obtain efficient algorithms, one of which is presented below.

## 9.2 Separation Method

The block structure of the quasi-static optimal power flow given by (9.3) entails the use of distributed optimization. Different approaches such as dual decomposition and/or alternating direction method of multipliers can be utilized in this case. However, we present a simple approach based on the dual for the balance constraint. First, we define two indicator functions:

$$\mathbf{I}_{\Omega_t} = \begin{cases} 0 & \{v, p\} \in \Omega_t \\ +\infty & \text{otherwise}, \end{cases} \tag{9.7}$$

$$\mathbf{I}_{\Gamma_t} = \begin{cases} 0 & \{s, e\} \in \Gamma_t, \\ +\infty & \text{otherwise}. \end{cases} \tag{9.8}$$

Then, we include the coupling constraint into the objective function via Lagrange multipliers, obtaining the following Lagrangian function:

$$\mathcal{L}(v, p, \lambda, s, e) = \sum_t \ell_t(v) + \lambda_t^\top (s_t - h_t(v, p)) + \mathbf{I}_{\Omega_t}(v, p) + \mathbf{I}_{\Gamma_t}(s, e), \tag{9.9}$$

where $\lambda$ represents the dual variables associated with (9.6). This problem can be solved in two stages, namely

$$\min_{v,p,\lambda} \left( \min_{s,e} \mathcal{L}(v, p, \lambda, s, e) \right). \tag{9.10}$$

The internal problem is labeled as the master problem and returns the decisions on the storage devices:

$$\{s, e\} \in \operatorname{argmin} \left( \sum_t \lambda_t^\top s_t + \mathbf{I}_{\Gamma_t}(s, e) \right). \tag{9.11}$$

Meanwhile, the external problem is divided into a sequence of independent optimal power flow problems:

$$\{v_t, p_t, \lambda_t\} \in \operatorname{argmin} \left( \ell_t(v) + \lambda_t^\top (s_t - h_t(v, p)) + \mathbf{I}_{\Omega_t}(v, p) \right). \tag{9.12}$$

Figure 9.1 shows the basic structure of the proposed methodology. First, the master problem is solved in a centralized manner. This problem returns values of $s_t$ and $e_t$ that are required by each of the instances of optimal power flow at the time $t$. These subsidiary problems, in turns, return a new set of dual variables $\lambda_t$ that are required by the master problem. The process continues until convergence is achieved.

Notice the model is a large-scale non-convex optimization. Therefore, we do not have guarantee or convergence or global optimality. However, the master problem is a simple linear programming problem, and the subproblems for each time $t$ are conventional optimal power flows that can be solved efficiently in practice via linearization or convexification. Notice also the great opportunity for parallelization, since each of the optimal power flow problems is independent among them.

Convergence can be improved if the master problem is replaced by the following proximal operator:

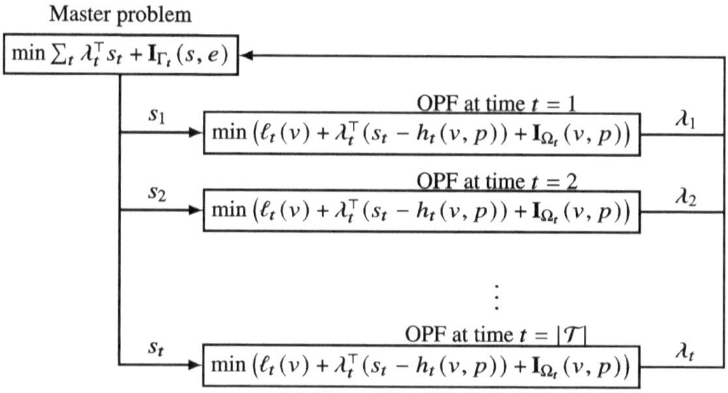


**Fig. 9.1** Schematic representation of the subproblems for the quasi-static optimal power flow in dc networks

$$\text{prox}_\alpha(\lambda, \sigma) = \text{argmin}\left(\sum_t \lambda_t^\top s_t + \mathbf{I}_{\Gamma_t}(s, e) + \frac{\alpha}{2}\|s - \sigma\|\right), \tag{9.13}$$

where $\alpha$ is a regularization parameter. The quadratic term in this operator allows the objective function to be strictly convex, making the solution unique and well defined. Fixed points of $\text{prox}_\alpha$ are minimizers of the original problem. Therefore, the convergence of the algorithm can be studied via fixed-point theory [6].

## 9.3 Numerical Results

A direct current meshed power distribution network presented in Fig. 9.2 is proposed to evaluate the methodology. The topology of the grid and the parameters of resistance and loads are based on the IEEE 33-node test system [7]. The substation is placed in Node 1, while solar photovoltaic systems are placed in Nodes $\{8, 12, 16, 20, 24, 32\}$; wind turbines are also considered in Nodes $\{4, 28\}$, as well as energy storage devices in Nodes $\{10, 20, 30\}$. We neglect the power loss of these storage devices.

For the sake of reproducibility and replicability, parameters of the grid are given in Table 9.1. In this table, $d$ represents the load, $\overline{p}$ the maximum generated power, $\overline{s}$ the maximum stored power, and $\overline{e}$ the maximum stored energy. The minimum power and energy is zero in all cases. The grid is not radial; then the nodal values are associated with the second column up to Node 33.

The profiles of the load curve and the available generation are depicted in Fig. 9.3. They consist of a 24-h horizon discretized in the order of the minutes to obtain 1440 scenarios, with their natural random behavior. Both solar and wind energy are considered. Energy prices change at different times of the day, being

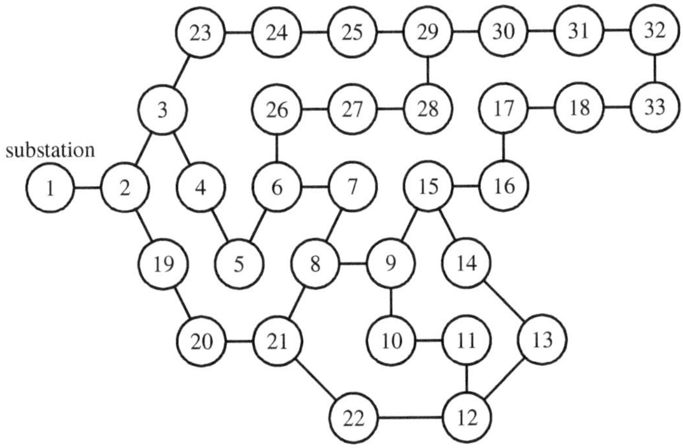


**Fig. 9.2** Proposed 33-node meshed DC distribution system

$c_t = 0.3 \forall t \in [0, 15]$, $c_t = 0.4 \forall t \in [15, 18]$, $c_t = 0.3 \forall t \in [18, 22]$. Therefore, power losses may have a different impact according to the time. The optimization process must decide not only about the usage of the storage devices but also about the power generated by each renewable resource. In some cases, the algorithm requires power curtailment.

The distributed algorithm was implemented in JuMP [8], an algebraic modeling language embedded in the Julia programming language [9]. The package for Interior-Point OPTimizer IPOPT [10] was used to solve both the master problem and the distributed optimal power flow problems. A computer with an Intel (R) Core(TM) i7-8700 CPU @ 3.20 GHz with x86-64 architecture was used. The total time calculation was 64.5 seconds, with 62.53 M allocations in 17.887 GiB.

Results for the generated power after quasi-dynamic time-series optimization are depicted in Fig. 9.4. In all cases, the generation tends to its maximum according to the availability of the primary resource. The power in the substation is highly reduced due to the combined effect of the power generation and the energy storage. In all cases, the profiles maintain a high variability. In practice, the quasi-static time-series resolution is often limited by the available measurement data resolution. However, a simplified discretization by hours would not capture this variability. This analysis can be performed for different time-series and different scenarios in order to obtain a more in-depth understanding of the grid performance. Some probabilities for each time-series scenario may be included into the model.

Although all the renewable resources receive the same available profile, according to the corresponding technology (e.g., solar or wind), the actual generated power varies slightly. This behavior is explained by the fact that the algorithm makes small changes in the generation to minimize power loss, according to the placement of the component in the grid. In those cases, there is some power curtailment.

Results for the storage devices are depicted in Fig. 9.5. The behavior of the energy storage devices is quite similar in all cases: They start charging in the

**Table 9.1** Parameters of the test system for the quasi-static optimal power flow

| From | To | $r$ (p.u.) | $\overline{p}_{km}$ (p.u.) | $d$ (p.u.) | $\overline{p}$ (p.u.) | $\overline{s}$ (p.u.) | $\overline{e}$ (p.u.) |
|---|---|---|---|---|---|---|---|
| 1 | 2 | 0.0006 | 0.2 | 0.100 | 0.0 | 0.0 | 0.0 |
| 2 | 3 | 0.0031 | 0.2 | 0.090 | 0.0 | 0.0 | 0.0 |
| 3 | 4 | 0.0023 | 0.2 | 0.120 | 0.4 | 0.0 | 0.0 |
| 4 | 5 | 0.0024 | 0.2 | 0.060 | 0.0 | 0.0 | 0.0 |
| 5 | 6 | 0.0051 | 0.2 | 0.060 | 0.0 | 0.0 | 0.0 |
| 6 | 7 | 0.0012 | 0.2 | 0.200 | 0.0 | 0.0 | 0.0 |
| 7 | 8 | 0.0107 | 0.2 | 0.200 | 0.8 | 0.0 | 0.0 |
| 8 | 9 | 0.0064 | 0.2 | 0.060 | 0.0 | 0.0 | 0.0 |
| 9 | 10 | 0.0065 | 0.2 | 0.060 | 0.0 | 0.5 | 180.0 |
| 10 | 11 | 0.0012 | 0.2 | 0.045 | 0.0 | 0.0 | 0.0 |
| 11 | 12 | 0.0023 | 0.2 | 0.060 | 0.8 | 0.0 | 0.0 |
| 12 | 13 | 0.0092 | 0.2 | 0.060 | 0.0 | 0.0 | 0.0 |
| 13 | 14 | 0.0034 | 0.2 | 0.120 | 0.0 | 0.0 | 0.0 |
| 14 | 15 | 0.0037 | 0.2 | 0.060 | 0.0 | 0.0 | 0.0 |
| 15 | 16 | 0.0047 | 0.2 | 0.060 | 0.8 | 0.0 | 0.0 |
| 16 | 17 | 0.0080 | 0.2 | 0.060 | 0.0 | 0.0 | 0.0 |
| 17 | 18 | 0.0046 | 0.2 | 0.090 | 0.0 | 0.0 | 0.0 |
| 2 | 19 | 0.0010 | 0.2 | 0.090 | 0.0 | 0.0 | 0.0 |
| 19 | 20 | 0.0094 | 0.2 | 0.090 | 0.8 | 0.6 | 170.0 |
| 20 | 21 | 0.0026 | 0.2 | 0.090 | 0.0 | 0.0 | 0.0 |
| 21 | 22 | 0.0044 | 0.2 | 0.090 | 0.0 | 0.0 | 0.0 |
| 3 | 23 | 0.0028 | 0.2 | 0.090 | 0.0 | 0.0 | 0.0 |
| 23 | 24 | 0.0056 | 0.2 | 0.420 | 0.8 | 0.0 | 0.0 |
| 24 | 25 | 0.0056 | 0.2 | 0.420 | 0.0 | 0.0 | 0.0 |
| 6 | 26 | 0.0013 | 0.2 | 0.060 | 0.0 | 0.0 | 0.0 |
| 26 | 27 | 0.0018 | 0.2 | 0.060 | 0.0 | 0.0 | 0.0 |
| 27 | 28 | 0.0066 | 0.2 | 0.060 | 0.5 | 0.0 | 0.0 |
| 28 | 29 | 0.0050 | 0.2 | 0.120 | 0.0 | 0.0 | 0.0 |
| 29 | 30 | 0.0032 | 0.2 | 0.200 | 0.0 | 0.7 | 190.0 |
| 30 | 31 | 0.0061 | 0.2 | 0.150 | 0.0 | 0.0 | 0.0 |
| 31 | 32 | 0.0019 | 0.2 | 0.210 | 0.8 | 0.0 | 0.0 |
| 32 | 33 | 0.0021 | 0.2 | 0.060 | 0.0 | 0.0 | 0.0 |
| 8 | 21 | 0.0125 | 0.2 | | | | |
| 9 | 15 | 0.0125 | 0.2 | | | | |
| 12 | 22 | 0.0125 | 0.2 | | | | |
| 18 | 33 | 0.0031 | 0.2 | | | | |
| 25 | 29 | 0.0031 | 0.2 | | | | |

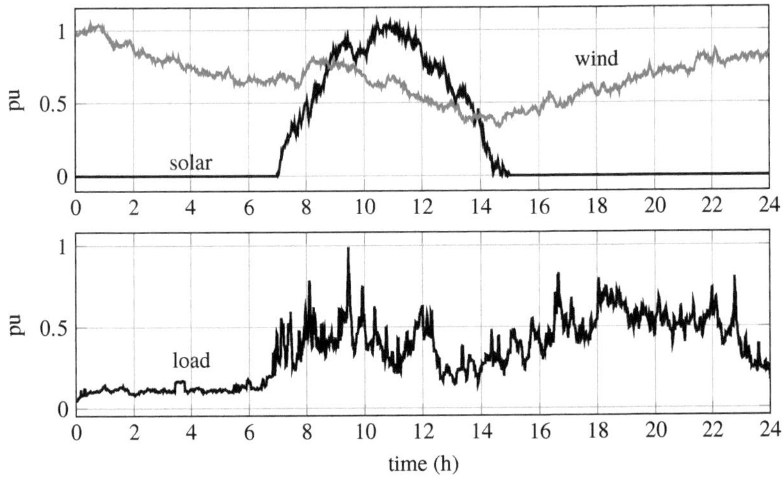


**Fig. 9.3** Profiles of load and generation for the quasi-static simulation

morning to discharge in the afternoon when the load and the price of electricity are higher.

## 9.4 Conclusions

The quasi-static optimal power flow allowed a better analysis of dc distribution networks with high penetration of renewable generation and energy storage devices. It was capable of optimizing the behavior of the grid in a time span considering both wind and solar generation. The storage devices were key to minimizing total losses, especially at peak hours when the solar generation is zero.

The main challenge of this problem is the fact that it is a large-scale optimization problem. However, it was possible to split the variables and constraints to obtain distributed algorithms that worked efficiently in practice. In particular, it was possible to obtain a master problem consisting of a linear programming problem and a series of subsidiary optimal power flow problems (one per each time step). This structure showed a high efficiency, allowing to obtain results with a suitable time calculation.

Distributed algorithms constitute a rich research area. The algorithm presented in this chapter was designed with simplicity in mind. However, more sophisticated algorithms such as the alternating direction method of multiplier may produce faster convergence. More research is required in this direction.

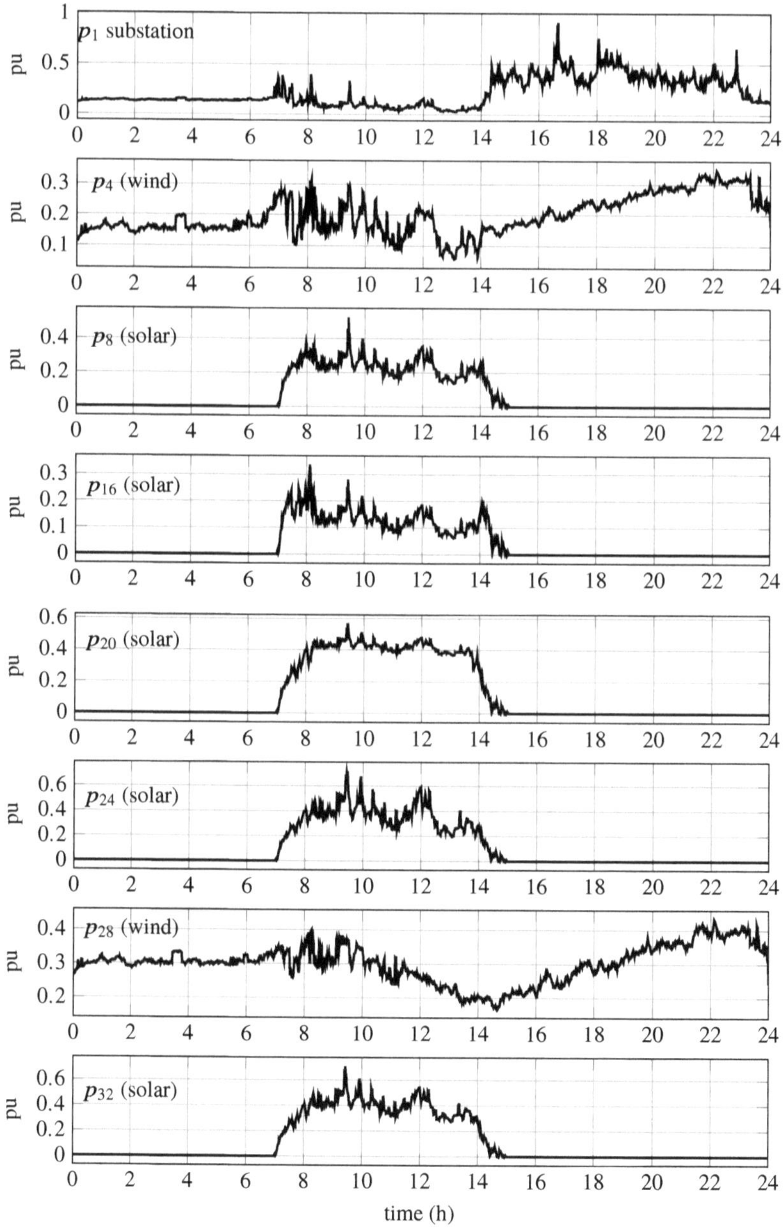


**Fig. 9.4** Profiles of actual generation resulting from the quasi-static optimal power flow

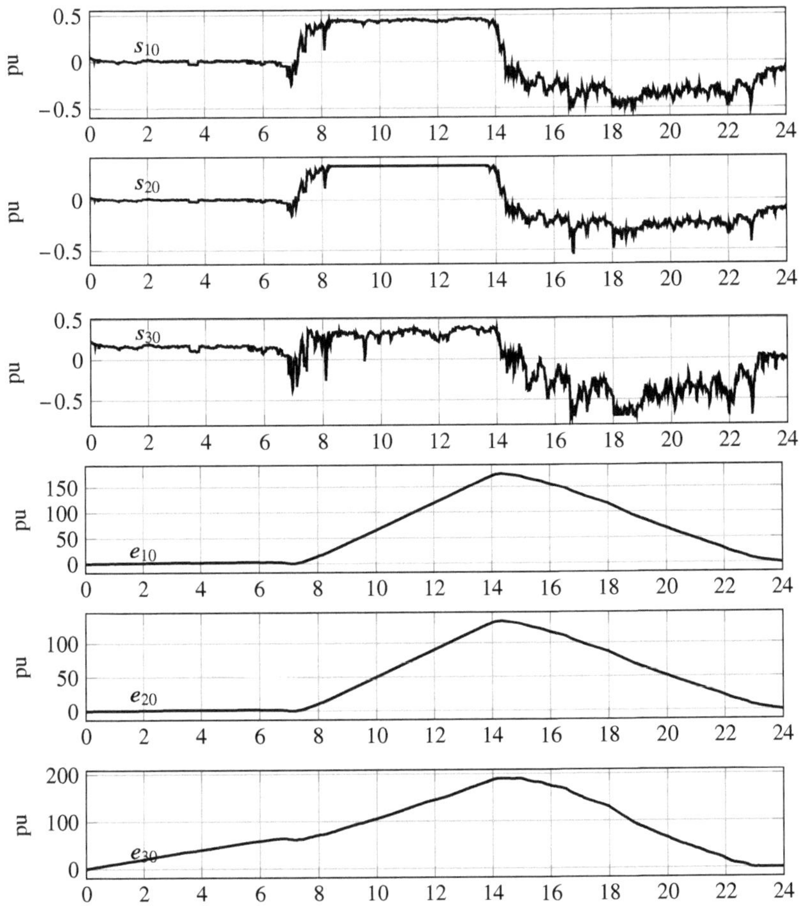


**Fig. 9.5** Profiles of storage devices

**Acknowledgments** This study was supported in part by project 6-24-2 Dinámica y control de elementos almacenadores de energía para el mejoramiento de la estabilidad en redes con alta penetración de recursos renovables basados en inversores (Dynamics and control of energy storage elements for improving the stability of networks with high penetration of inverter-based renewable resources), funded by the Vicerrectoría de Investigaciones, Innovación y Extensión (Office of Research, Innovation and Extension), Universidad Tecnológica de Pereira, Pereira, Colombia.

## References

1. Boyd, S., Parikh, N., Chu, E., Peleato, B., & Eckstein, J. (2011).
2. Taylor, J., & Peppanen, J. (2023). Distributed energy resource benchmark models for quasi-static time-series power flow simulations. In *CIGRE Technical Brochures Active distribution systems and distributed energy resources (C6).*

3. Garcés-Ruiz, A., Gil-González, W., & Montoya, O. D. (2025). Quasi-dynamic optimal power sharing in isolated dc microgrids. *Electric Power Systems Research, 239*, 111222.
4. Lukian, L., Anton, S., Pavlo, H., Anna, P., & Vsevolod, P. (2024). *Development and utilization of a quasi-dynamic model for power system analysis* (pp. 1–29). Springer Nature Switzerland.
5. Marujo, D., & Sousa Neto, A. A. (2024). AnaDyn: Educational tool for dynamic and quasi-steady-state simulations of electrical power systems. *Electrical Engineering, 106*(3), 3321–3335.
6. Granas, A., & Dugundji, J. (2013). *Fixed point theory*. Springer.
7. Dolatabadi, S. H., Ghorbanian, M., Siano, P., & Hatziargyriou, N. D. (2021). An enhanced IEEE 33 bus benchmark test system for distribution system studies. *IEEE Transactions on Power Systems, 36*(3), 2565–2572.
8. Lubin, M., Dowson, O., Dias Garcia, J., Huchette, J., Legat, B., & Vielma, J. P. (2023). JuMP 1.0: Recent improvements to a modeling language for mathematical optimization. *Mathematical Programming Computation, 15*(3), 581–589.
9. Bezanson, J., Edelman, A., Karpinski, S., & Shah, V. B. (2017). Julia: A fresh approach to numerical computing. *SIAM Review, 59*(1), 65–98.
10. Wächter, A., & Biegler, L. T. (2006). On the implementation of an interior-point filter line-search algorithm for large-scale nonlinear programming. *Mathematical Programming, 106*(1), 25–57.

# Index



© The Editor(s) (if applicable) and The Author(s), under exclusive license to Springer Nature Switzerland AG 2026

O. D. Montoya Giraldo et al., *Optimal Power Flow Methods for DC Networks*, Power Systems, https://doi.org/10.1007/978-3-032-05734-1

**S**



**T**



**V**

MIX
Papier aus verantwortungsvollen Quellen
Paper from responsible sources
FSC® C105338

If you have any concerns about our products,
you can contact us on
**ProductSafety@springernature.com**

In case Publisher is established outside the EU,
the EU authorized representative is:
**Springer Nature Customer Service Center GmbH**
**Europaplatz 3, 69115 Heidelberg, Germany**

Printed by Libri Plureos GmbH
in Hamburg, Germany